Biofuel Production

Biofuels and bioenergy have emerged as an alternative option based on their sustainability, concomitant waste treatment, and site-specific flexibility. This book encompasses all the knowhow of different biofuel production processes through biological methods. It describes recent advancements in all major biofuel technologies such as biohydrogen, biomethane, bioethanol, syngas and so forth. Related protocols supported by schematic representation are included, encompassing comprehensive up-to-date scientific and technological information in biofuels and bioenergy.

Features:

- Includes practical approaches focused on process design and analysis in biofuel production via biological routes
- Discusses kinetic equations of different microbial systems
- Provides comprehensive coverage of biochemical kinetics and equations related to biofuel process
- Describes protocols for setting up of experiments for pertinent biofuel technologies
- Emphasis on practical engineering approaches and experiments

This book is aimed at researchers and graduate students in chemical, biochemical and bioprocess engineering, and biofuels.

Biofuel Production

Biological Technologies and Methodologies

Ramkrishna Sen and Shantonu Roy

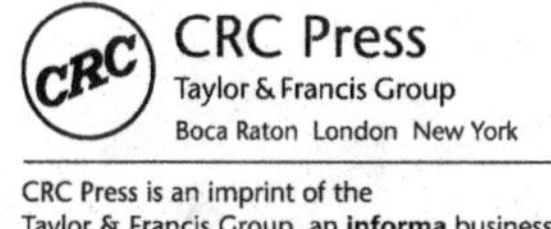

CRC Press is an imprint of the
Taylor & Francis Group, an informa business

Designed cover image: ©Shutterstock

First edition published 2023
by CRC Press
6000 Broken Sound Parkway NW, Suite 300, Boca Raton, FL 33487-2742

and by CRC Press
4 Park Square, Milton Park, Abingdon, Oxon, OX14 4RN

CRC Press is an imprint of Taylor & Francis Group, LLC

© 2023 Ramkrishna Sen and Shantonu Roy

Reasonable efforts have been made to publish reliable data and information, but the author and publisher cannot assume responsibility for the validity of all materials or the consequences of their use. The authors and publishers have attempted to trace the copyright holders of all material reproduced in this publication and apologize to copyright holders if permission to publish in this form has not been obtained. If any copyright material has not been acknowledged please write and let us know so we may rectify in any future reprint.

Except as permitted under U.S. Copyright Law, no part of this book may be reprinted, reproduced, transmitted, or utilized in any form by any electronic, mechanical, or other means, now known or hereafter invented, including photocopying, microfilming, and recording, or in any information storage or retrieval system, without written permission from the publishers.

For permission to photocopy or use material electronically from this work, access www.copyright.com or contact the Copyright Clearance Center, Inc. (CCC), 222 Rosewood Drive, Danvers, MA 01923, 978-750-8400. For works that are not available on CCC please contact mpkbookspermissions@tandf.co.uk

Trademark notice: Product or corporate names may be trademarks or registered trademarks and are used only for identification and explanation without intent to infringe.

ISBN: 9781032124452 (hbk)
ISBN: 9781032124483 (pbk)
ISBN: 9781003224587 (ebk)

DOI: 10.1201/9781003224587

Typeset in Times
by Newgen Publishing UK

Contents

Author Biographies

Ramkrishna Sen is presently professor and head of the Department of Biotechnology; chairperson of the School of Bioscience; chairperson of the Central Research Facility (Life Science Division), and professor-in-charge at the Advanced Lab for Plant Genetic Engineering and Joint Faculty of the P K Sinha Center for Bioenergy & Renewables at IIT Kharagpur. He served as a Fulbright Visiting Professor in the Columbia University, New York, USA. Professor Sen has been engaged in R&D activities in the areas of energy, environment, water and healthcare, with a major focus on green process and product development in microalgal-and-microbial biorefinery models.

Shantonu Roy is an assistant professor in the School of Community Science and Technology at the Indian Institute of Engineering Science and Technology Shibpur, West Bengal, India. He was a gold medalist at undergraduate microbiology level (Vidyasagar University) and completed his master's degree in biotechnology at Jadavpur University, Kolkata. He then moved to the Indian Institute of Technology (IIT) Kharagpur, West Bengal, India where he pursued his PhD. He has been awarded the prestigious Indian government Department of Science and Technology (DST) INSPIRE faculty, and early career awards. He is among the top 20 scientists working in the field of biohydrogen in India (according to the DST India Country Status Report on Hydrogen and fuel cells, 2020). His research interests focus on hydrogen production and methane production from thermophilic bacteria. He has developed expertise on metagenomic analysis of thermophilic mixed consortia, proteomics and strain improvement by mutagenesis.

Preface

The evolution of human civilization went through a rapid modernization which has been catalyzed by the use of various kinds/types of energy sources. The dependency of the human population on fossil fuels such as coal, natural gas, and petroleum has impacted the global weather scenario. The emission of greenhouse gases associated with the use of fossil fuels has led to the alteration of global atmospheric temperatures. This has not only impacted weather patterns but also lead to rising oceanic water levels, leading to future submergence of the landmass.

As there is a substantial growth in transportation fuel demand; a gradual increase in fossil fuel price has increased the economic burden on third world countries. Moreover, the proven reserves of fossil fuel reserves are limited. Therefore, a recent impetus has been given in the field of renewable energy generation, keeping in mind the need of future generations. Various types of renewable energy sources viz. solar, hydroelectric, biomass gasification, and biofuels, are explored. Biofuels and bioenergy have emerged as a potential option based on their sustainability, concomitant waste treatment, and site-specific flexibility.

Bioenergy such as bioethanol, biobutanol, biohydrogen, biomethane and bioelectricity from the microbial fuel cell, are a few technologies that are studied extensively. Biofuels can be truly considered as renewable energy only if the feedstock used for their production is also renewable.

The technologies involved in the production of biofuels involves the interface between various disciplines of science and engineering. The best example of it is bioethanol production which needs a knowledge of microbiology, chemical engineering, agriculture, and food technology.

Therefore, this book is written for budding researchers who intend to start their career in the above-mentioned bioenergy technologies. It gives a comprehensive background of the biofuel technologies such as microbial bioprocess engineering, enzymology, chemical catalytic methods. The major emphasis is on the illustration of basic protocols involved in the production of these biofuels. This would help the researchers, graduate students, faculty, and industrial research and development units to refer to this book for starting any experiments related to biofuel-based technologies.

Chapter 1 gives brief information regarding the topics included in this book. It includes basic definitions, the global scenario regarding the booming energy crisis, and discusses the role of bioenergy in sustainability and augmenting the energy demand. It also discusses the environmental and ecological impact of biofuel technologies.

Chapter 2 gives insight into the biohydrogen production process. It includes basic microbial diversity involved in the process and their biochemical uniqueness. It also provides a lucid explanation of the experimental procedures involved in facultative and obligate anaerobic biohydrogen production. The challenges of cultivating thermophilic microorganisms for biohydrogen production and experimental setup are also discussed.

Chapter 3 gives insight into the biomethane production process. It includes basic microbial diversity involved in the process and their biochemical uniqueness. It also provides a lucid explanation of the experimental procedures involved in facultative and obligate anaerobic biohydrogen production. The antagonistic microbial dynamics which influence the overall efficiency of the process has been discussed in detail. The challenges of cultivating thermophilic microorganisms for biomethane production and experimental setup are also discussed.

Chapter 4 encompasses all the available pathways of bioethanol production. Moreover, the experimental protocol for the same has also been elucidated.

Chapter 5 deals with the microbial and bioprocess aspects of biobutanol production. It also describes the challenges of cultivating biobutanol producing microbes under reduced partial pressure. The protocols for setting experiments for biobutanol production is discussed extensively.

Chapter 6 encompasses the scientific and technological details of oil production from oleaginous microorganisms such as algae and yeasts. The process of cultivation of photosynthetic algae in different reactor configurations is also discussed in detail. Challenges of the production of bio-oil from yeasts over algae and its downstream processing are discussed in detail. The strategies of microorganism biomass harvesting in conventional ways and its application has also been elucidated.

Chapter 7 discusses the prospect of exploiting electrogenic bacteria for electricity production. The microbial diversity related to bioelectricity production and their respective biochemical pathways is discussed. Moreover, how to fabricate/operate a basic dual chamber microbial fuel cell (MFC) and single chamber MFC is explained. This book covers in-depth knowledge about an interdisciplinary field of biofuel production. It can be ideally used as a reference book for a three-credit-hour semester course in bioenergy and biofuels. Furthermore, this book will encompass comprehensive up-to-date scientific and technological information for researchers working in the field of biofuels and bioenergy.

1 Introduction to Biofuel and Bioenergy

1.1 INTRODUCTION

In the Paleolithic era, when human evolution was in its infancy, the discovery of energy sources such as fire, animal fats, and plant resins facilitated its anthropogenic development. By the time we reached the Neolithic era, the agricultural revolution had started and promoted the early civilization to shun the Bedouin lifestyle, leading to permanent settlements. It can be safely assumed that availability of potable water sources, food, and primary energy sources were instrumental in developing a functional civilization. The earliest evidence of civilization was found in Mesopotamia, the Indus valley, and Egypt, dated around 4000 BC. As the civilization further flourished, it was the discovery of coal which turned the tide in favor of rapid industrial growth and the concomitant development of human civilization. The earliest evidence of coal usage, by Romans, was found in England (AD 100–200) (Dearne and Branigan, 1995). However, it was an industrial revolution in the mid-1800s, which exploited the most actual potential of coal as a reliable fuel source. Another fossil fuel that was gaining importance during this time was crude petroleum oil. With the advent of refining technologies in the early 1900s, the hidden potential of crude oil was realized, be it petroleum, diesel, lubricants, or multi-utility products such as plastics. Extensive use of petroleum has propelled human civilizational development rapidly (Black and Brian, 2014).

The extensive and non-judicious usage of fossil fuels has given rise to many environmental hazards. Combustion of fossil fuels leads to the production of greenhouse gases such as CO_2, CO, CH_4, and H_2O, etc. These greenhouse gases are essential for maintaining the optimum temperature of the earth to sustain life. However, the excess accumulation of these gases has led to global temperature increment (Nayak et al., 2014). This steady increment in global temperature has impacted the global weather pattern. Increment in global temperature leads to the melting of ice at polar regions, which results in flash floods and submergence of low-lying lands. If corrective measures are not implemented on a war footing, human civilization could face an existential crisis soon. Even the proven reserves of fossil fuels are now staring at imminent exhaustion. The need of the hour is to find an alternative to a conventional source of energy. In this regard, we as

DOI: 10.1201/9781003224587-1

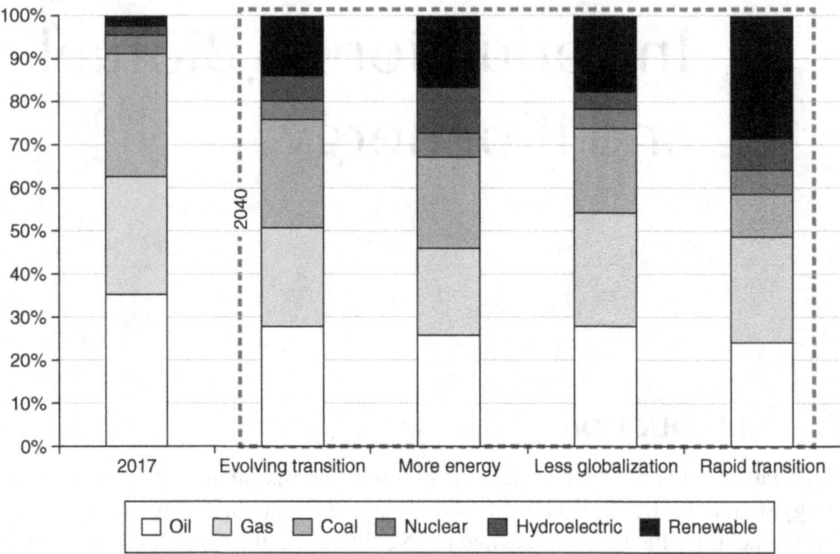

FIGURE 1.1 Prospective trend of contribution by renewable energy resources towards energy demands.

Source: (BP energy outlook, 2019).

humans have taken significant strides towards exploring renewable, carbon-neutral energy sources. In the past decades, technologies such as photovoltaic solar cells, hydrogen fuel cells, bio-CNG, and biofuels, have been explored extensively (Das and Roy, 2016).

According to a British petroleum report, by 2040 the pattern of energy supply will witness a substantial increase in the contribution of renewables (Figure 1.1). This report projects different scenarios of energy supply viz. evolving transition, less globalization, and rapid transition. It has postulated that under rapid transition towards cleaner energy would contribute about 28% of the total demand compared to just 2% in 2017.

Suppose we see the projection of accessibility to electricity under a sustainable development scenario from 2010 to 2030. A developing Asia and other developing countries would substantially improve their electricity supply chain post-2020. The Middle East would witness a spike in electricity generation post-2020, as increased dependence on solar-based electricity is envisaged. Thereby, soon a decreasing dependency will be observed on limited oil reserves present in these countries. Quenching the thirst for clean and reliable energy soon can only be realized by investing more resources towards developing renewable technologies.

1.2 RENEWABLE ENERGY SCENARIO

Various kinds of energy sources have fueled the anthropogenic and technological development of the last century. Rapid innovation in the field of alternative energy,

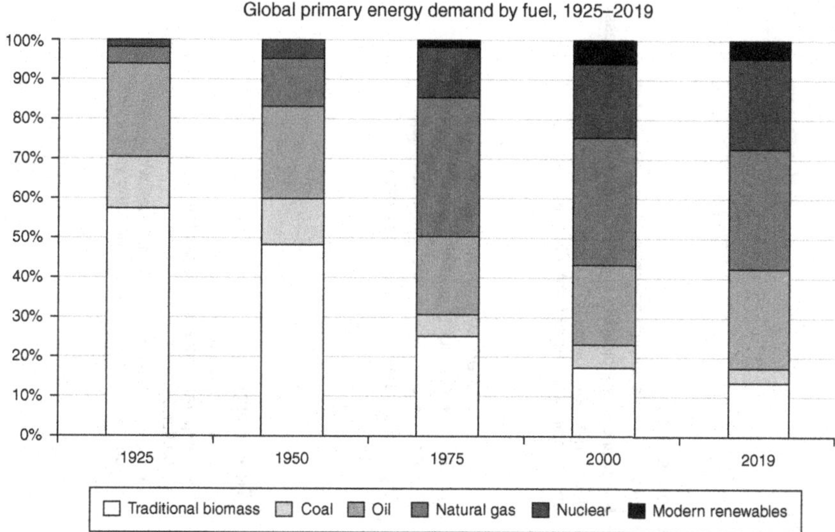

FIGURE 1.2 Systemic change observed in types of energy resources supplying global energy demand.

Source: (Energy Technology Prospective 2020).

storage, and power management has paved the path for the development of nuclear power, non-hydro renewable energies, and integrated energy systems. The early wave of innovation in renewable energy cropped its head in the 20th century and further expanded significantly (Figure 1.2).

Coal and oil, which propelled the industrial growth during the 18th and 19th centuries, started flowing a declining trend in the 21st century. This observation corroborates with a steady increase in the contribution of renewable energy towards global energy demand. Since 2000, there has been a steady increment of investment in renewable energy for power generation by Europe, the United States, China, and India. A sudden transition from the usage of conventional sources of energy to renewable energy is not feasible. The current humungous energy demand cannot be fulfilled with the available renewable energy technologies. Instead, an augmented system can be developed where renewable energy will supplement the conventional energy sources. Gradually, with the development of infrastructure and policies, they will eventually replace conventional sources.

1.2.1 Technologies Available for Renewable Energy Generation

The geological availability of fossil fuels is not evenly distributed throughout the globe, and therefore, the countries rich in these resources do mark their influence in global geopolitics. However, in the case of renewable energy, the distribution of the resources is comparatively more dispersed than fossil fuels. Various renewable energy sources such as:

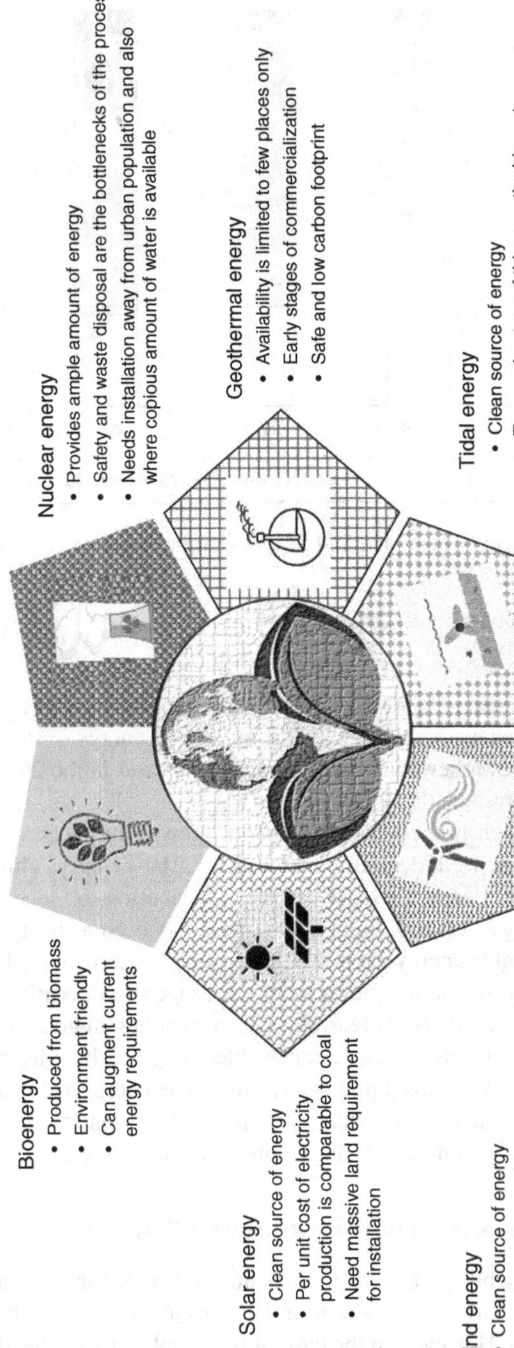

Renewable energy sources
Technologies available
Bioenergy
• Produced from biomass
• Environment friendly
• Can augment current energy requirements
Solar energy
• Clean source of energy
• Per unit cost of electricity production is comparable to coal
• Need massive land requirement for installation
Wind energy
• Clean source of energy
• Need massive land requirement for installation
• Can be installed in particular area with consistent wind flow throughout the year
• Interferes with radar of private and military aviation
Nuclear energy
• Provides ample amount of energy
• Safety and waste disposal are the bottlenecks of the process
• Needs installation away from urban population and also where copious amount of water is available
Geothermal energy
• Availability is limited to few places only
• Early stages of commercialization
• Safe and low carbon footprint
Tidal energy
• Clean source of energy
• Temporal nature of tides are the biggest bottleneck
• Can be installed in particular area with consistent tidal flow throughout the year

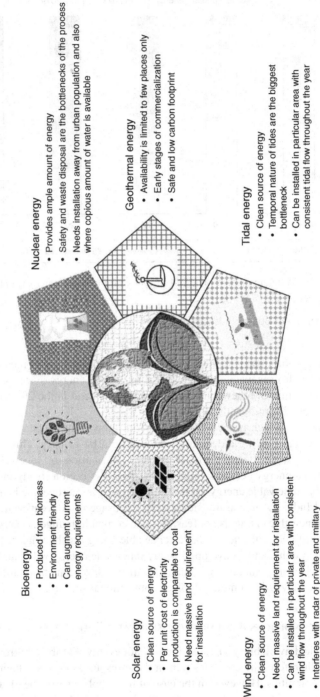

FIGURE 1.3 Types of renewable energy resources.

- Solar energy
- Wind energy
- Tidal energy
- Geothermal energy
- Nuclear energy
- Bioenergy

have shown their exploitability and commercial feasibility in the recent decade (Figure 1.3).

1.2.1.1 Solar-based Energy Generation

Solar energy is the most abundant resource and has been harvested successfully for energy generation. The solar radiation at the equatorial region receives 1.366 W/m^2 of energy, also known as the solar constant (Brock et al., 2012). If we consider this solar constant as 100%, then 19% of this gets absorbed by the atmosphere, which keeps our planet warm and habitable, and35% of the incidental energy gets reflected by clouds and dust. Solar energy is harvested by various technologies such as photovoltaic cells-based electricity generation, solar energy mediated, molten salt-based pressurized steam mediated electricity, and photo-catalytic energy generation (Prieto et al., 2016).

Various operational challenges are associated with solar energy viz.

- High installation price
- Requirement of land
- The complexity of solar panel fabrication
- Storage of electricity in batteries which have poor energy efficiency
- Poor recycling potential for solar panels
- Seasonal variation in the availability of solar energy

Various technological advancements are being implemented to mitigate the bottlenecks mentioned above. Recently, Perovskite solar cells have gained much importance due to their organo-inorganic ultrastructure, which has higher solar energy conversion efficiency than conventional solar cells (Assadi et al., 2018). The advent of Li-ion batteries and other solid-state metal batteries has improved the energy storage efficiency of solar cells. These batteries are of high energy density compared to conventional lead acetate/lead sulfuric acid batteries (Kim et al., 2020).

1.2.1.2 Wind Energy

The heating up of atmospheric air due to solar radiation causes the hot air to rise due to its lower density, creating a vacuum. This vacuum eventually gets filled with nearby cooler air; this leads to the generation of wind. The flow of air due to this phenomenon can be garnered for electricity generation. Installation of windmills can convert the mechanical energy of wind to electrical energy. The technology related to such a concept is relatively simple but is marred with various operational challenges such as:

- Non-availability of consistent wind flow throughout the year
- Disturbance to radars of the civil and military installation
- Enormous land requirement
- High installation cost

With the advent of sophisticated distributed energy management systems, the integration and exploitation of wind energy can be realized with its utmost potential.

In recent times, wind energy has become quite competitive, and the cost of electricity production also came down drastically (Quintana-Rojo et al., 2020). The feasibility of wind energy has been studied, based on five categories (de Oliveira Azevêdo et al., 2020):

a) Location of the turbine, surface roughness
b) Economic feasibility such as installation cost operation and maintenance cost, depreciation, etc.
c) A climatic condition such as wind speed, air density, pressure drop, temperature variation, etc.
d) Technical factors such as the height of turbine motor diameter, operation time, etc.
e) Political implications such as taxes, interest rate, payback period, policies, financial support, etc.

1.2.1.3 Tidal Energy

Tides are generated due to the gravitational pull exerted by the sun and moon over the oceans. The forces responsible for the generation of tides have an inverse relationship between the distance between two bodies In contrast, gravitational forces vary inversely to the square of the distance between objects (Thurman and Elizabeth, 1997).

Electricity generation using tidal energy has been considered as green energy in the form of zero-carbon footprint. Installation of tidal energy harvesting does not take much of the space compared with solar and wind farms. The world's largest solar park is spread across 45 km^2 in Rajasthan, India. In contrast, the largest tidal form takes around 12.5 km for generating 254 MW of energy (Cho et al., 2012). Another advantage of tidal energy is its predictable and continuous availability. The pattern of tides is cyclical, and therefore, it is easier to harness a reliable amount of energy out of it. Whereas, if we compare the same with wind flow, it is not at all consistent. According to a report published by Bloomberg, the UK saw zero electricity production from wind for nine days in just one month. (Morison, 2018).

Tidal energy can be harvested at relatively low speeds compared with wind energy. Since water has a density ten thousand times greater than air, the tidal waves at speeds as low as 1 m s^{-1} can generate electricity, whereas, wind speed of at least 3 to 4 m s^{-1} is required for the generation of electricity. The impact of tidal turbines and barrages on marine life is not clear and needs a more significant investigation. Installation of tidal energy infrastructure may interfere with the electromagnetic fields (EMF) of the marine ecosystems. This distribution of EMF and its impact on aquatic flora and fauna is not known at this moment (Barber et al., 2020). Another disadvantage is the high cost of installation and maintenance. Installing concrete barrages against oceanic waves is a great challenge. However, in the near future, with the advancement of technologies related to installation, tidal energy can certainly be realized to its most actual potential.

1.2.1.4 Geothermal Energy

The sub-surface of the earth has entrapped thermal energy. This thermal energy is carried to the surface of the earth with the help of water or steam. High or medium temperature is required for electricity generation. Such high-temperature zones are available near tectonically active regions. Countries like New Zealand, Iceland, Kenya, and El-Salvador are generating a significant amount of electricity using geothermal energy. The major advantage of geothermal energy is its consistent availability and not being influenced by climatic conditions, thus it can supply baseload and ancillary supplies. The emission profile of geothermal energy is quite low when compared with conventional fossil fuels.

Moreover, the operational efficiency of the geothermal heat pump is 25 to 50% greater than the conventional heat pump. It requires less space for installation. The geothermal energy harvesting systems have very few movable parts, therefore, the overall maintenance cost is low (Tester et al., 2006).

Despite its vast potential, there are chances that the notified location where geothermal energy is prominent might cool down after some point in time. This could lead to wastage of investment and capacity build-up. Moreover, the installation cost is also high, with a payback period of over ten years. There are various types of geothermal technology which are available for electricity generation. The utility of these technologies depends upon the climatic conditions, soil, and available land for installation. There is a ground loop system through which geothermal energy is trapped.

The ground loop is an assembly of series of pipes embedded deep inside the sub-surface geothermal zone. It serves as an interface for exchanging heat and thereby helps in the generation of steam. This super-heated steam is pressurized and used for electricity generation. It can be further categorized as an open ground loop and close ground loop system.

Open categorized loop system pulls out water from an open pond with the help of heat exchanger transfer the heat of it generate steam for electricity generation. The closed-loop geothermal system uses a fluid-filled pipe system which is buried 6–7 feet deep into the geothermal zone, and the hot fluid then exchanges using a heat exchanger to generate steam for electricity generation. In the coming few decades, geothermal energy could gain prominence among all available renewable energy sources.

1.2.1.5 Nuclear Energy

Nuclear energy is one of the most successful zero-carbon technologies which have been exhibited in recent years extensively. Nuclear energy is derived primarily from a nuclear fission reaction. In this, the U_{235} atom is bombarded with a fast-moving neutron, which leads to the formation of U_{236}. The U_{236} then further undergoes decay and generates Barium-144, Krypton-89, and 3 neutron particles. During this decay, an enormous amount of energy is liberated in terms of heat. This energy is harvested using liquid sodium or heavy water, which eventually helps in the generation of high-pressure steam. This high-pressure steam is then further utilized for electricity generation.

In the US, nuclear energy is the largest source of clean energy. It contributes about 55% of the nation's electricity demand. This helps in avoiding 470 million metric tons of CO_2 emission, which is equivalent to removing 100 million cars off the road (ONE, 2021). Globally, nuclear energy contributes about 25% of total energy need with electricity produced for 0.025 to 0.07 USD per kilowatt-hour.

Various technologies have been developed for harvesting nuclear fission. Based on the different types of reactors used for performing nuclear fission, the efficiency of energy generation varies (Dagen et al., 2020). There are primarily six different types of reactor systems that are operational (Table 1.1).

Recently, India has made ground-breaking progress towards developing a thorium-based nuclear reactor (Nayak et al., 2019). Thorium is available in abundance when compared to Uranium. Thorium-based technology involves the use of Thorium- 232-Uranium- 233 fueled reactors. This would park the way for the development of a self-sustained nuclear energy generating system.

The significant bottlenecks of nuclear energy are as follows:

a) Though nuclear energy is clean and has a shallow carbon footprint, it substantially impacts the environment. The significant impact comes from the mining of radioactive ores of Uranium and Plutonium which might contaminate local water bodies and can cause a detrimental effect on flora and fauna. The water discharged from the heat exchanger of the power is planned to get mixed with oceanic water leading to local temperature

TABLE 1.1
Different Types of Nuclear Reactor Configurations

Reactor	Fuel	Coolant	Moderator
Pressurized water reactor	Enriched UO_2	Water	Water
Boiling water reactor	Enriched UO_2	Water	Water
Pressurized heavy water reactor	Natural UO_2	Heavy Water	Heavy Water
Advanced gas-cooled reactor	Natural U, Enriched UO_2	CO_2	Graphite
Light water graphite reactor	Enriched UO_2	Water	Graphite
Fast neutron reactor	PO_2 and UO_2	Liquid Sodium	None

increment. This may cause damage to the aquatic flora and fauna of that ecological niche.

b) Disposal of nuclear-spent fuel is a big challenge. The spent fuels are enclosed in implies and buried under the earth or the oceanic beds.
c) It is a water-intensive process and needs a consistent source of water supply. The United States consumes around 320 billion gallons of water to produce nuclear power.
d) Changes of a mishap resulting in the nuclear holocaust are a looming danger associated with nuclear energy.

1.2.1.6 Bioenergy and Biofuel from Biomass

The last two decades witnessed the emergence of bioenergy as technological advancement in this field gained greater emphasis. Bioenergy has been envisioned as one of the most promising renewable energy sources, as the feedstock for such technologies is renewable. If we consider the total energy demand of 2019, bioenergy contributes about 9% followed by nuclear power (5%), hydropower (3%), solar, and wind (1%) (Popp et al., 2020).

Bioenergy can be defined as any energy that is derived from biomass. At the same time, biofuels can be defined as fuel that is generated through the metabolic activity of microorganisms utilizing biomass residues or their processed derivatives as feedstock. In terms of utility, biomass has been traditionally used for power generation, heating homes, and transportation. Whereas liquid biofuels such as ethanol, biodiesel, and biobutanol have generally been used for transportation only.

Photosynthetic autotrophs tap the solar energy and sequester atmospheric CO_2 into complex organic biomass. Bioenergy or biofuels, when burned for energy, eventually release back the sequestered CO_2 to the atmosphere. In this way, bioenergy or biofuel can be considered as a carbon-neutral fuel. So, by maintaining a delicate balance between cultivating biomass and subsequently using them, we can develop a renewable carbon-neutral and sustainable process.

To realize the goal of implementing a biofuel and bioenergy-based economy, agricultural reforms would be required in the near future. Contract farming, genetically modified plants, drought-resistant food crops, etc., provide a sustainable supply of biomass for bioenergy and biofuel production. Three prospecting scenarios have been mooted by a joint study of the US Department of Energy and the US Department of Agriculture (Perlack et al., 2005).

- Scenario 1: Sustainable availability of biomass feedstocks from current cultivatable agricultural lands.
- Scenario 2: Implementation of various technologies to improve biomass productivity from conventional crops.
- Scenario 3: Through the implementation of technological advancement in conventional crop and new perineal crop cultivation, thereby increasing biomass availability.

The biomass, which is available at present, represents the baseline availability, which encompasses sustainable biomass residues, tillage practices, residue collection methodologies and practices, and secondary and tertiary residues (Perlack et al., 2005).

The use of biomass as feedstock for bioenergy and biofuel production includes the interplay of various mechanical, chemical, and biochemical process treatments. Biomass can be converted into solid, liquid, and gaseous fuel depending upon the technologies implemented. Biomass can be dried and pelletized for bulk handling, which further can be used as solid fuel. A high-pressure-driven pelletizer is used for the production of biomass pellets of 0.8 cm in size. Biomass pellets as fuels have given new hope to third-world countries that do not have access to clean energy resources (Zang et al., 2018).

1.2.1.6.1 Technologies Available to use Biomass for Bioenergy Generation

- Torrefaction
- Hydrothermal carbonization
- Biomass to syngas

Torrefaction: The biomass is dried and heated in torrefaction at 200–300°C under low oxygen concentration and near atmospheric pressure. The biomass undergoes degradation, leading to the release of various volatile gases. The loss of energy is also observed with the release of these volatile gases. The extent of mass and energy yield from raw biomass and terrified biomass depends upon the process parameters viz rate of heating, reaction temperature, time of heating, and type of biomass. During the torrefaction process, the mass loss is greater than the energy loss due to the release of gaseous products. This results in the formation of denser energy-containing carbonized biomass. Biomass is a conglomeration of various biological polymers such as cellulose, hemicellulose, lignin, and resin. The reaction kinetics for the decomposition of these polymers depends upon the process temperature. The order of reactivity is hemicellulose > lignin > cellulose. The hemicellulose undergoes limited carbonization and de-volatilization at a temperature below 250°C. On further increasing temperature above 250°C, the hemicellulose undergoes rapid customization and de-volatilization., whereas cellulose is quite thermostable and undergoes suboptimal carbonization at 250°C. On heating, the biomass loses its tenacious property due to the breakdown of the intricate network present between hemicellulose with cellulose. This results in the depolymerization of biomass. In a case study conducted with willow biomass, torrefaction showed five main reaction products viz, solid, lipid, organics, gases, and water, along with their energy yields (Table 1.2).

The data of Table 1.2 shows the predominant contribution of solid reaction product (terrified biomass) towards mass yield and energy yield. Water showed the second-highest mass yield, and it originates from the loss of moisture from all the polymers. The majority of energy loss from solid biomass is contributed by the organic and lipid content of the biomass. During the torrefaction process, the

TABLE 1.2
Reaction Product from Torrefaction of Willow and Respective Energy Yield

Reaction products	Mass yield (%)	Energy yield (LHV) %
Solid	8.75	94.90
Lipids	1.40	3.40
Organics	1.70	1.60
Gases	1.40	0.10
Water	8.00	0.00

Source: Bergman and Ki el., 2005.

biomass loses a relatively higher amount of oxygen and hydrogen compared to carbon. The organic fraction generally contains furans, short-chain fatty acids, and aromatic compounds. The gaseous products generally contain CO_2 and CO and a considerable amount of oxygen. Due to the release of these organic and gaseous products, the solid residue gains more excellent calorific value than raw biomass. The torrefaction process can generate solid biomass with lower heating values ranging from 19 to 23 MJ kg^{-1}. The torrefied biomass is the least hygroscopic and undergoes some internal chemical transformation. Such biomass is also resistant to biological degradation. During this process, biomass becomes more porous, which results in a decrease in volumetric density (180–300 Kg/m^3). The torrefied biomass can be further densified by a technique called pelletization. High-density torrefied pellets can be produced through pelletization with a range of 700–900 Kg/m^3 (Bergman, 2005).

Hydrothermal carbonization: Another technology that gathered interest in recent years is the hydrothermal carbonization of biomass. The hydrothermal carbonization process is also called artificial coalification. The significant advantage of this process is that it can process wet biomass is subjected to high temperature and pressures under which biomass undergoes thermos-chemical carbonification. The process temperature is kept in the region of 180°C to 280°C, and pressure is kept in a range of (2–10 MPa). The entire process of wet torrefaction occurs within a time frame of 10 to 20 min (Lu et al., 2009). At 200°C to 250°C, the water's dielectric constant decreases and behaves as a non-polar solvent as well as an acidic solvent. Under these conditions, the hemicellulose starts hydrolyzing whereas, cellulose is sturdier and starts hydrolyzing at 230°C. Lignin starts degradation at a temperature above 260°C and leads to the formation of phenol and phenolic derivatives. The inorganic fraction of biomass remains unchanged throughout the HTC process.

Biomass to syngas: It is a promising technology in which the biomass is converted into gaseous products such as CO, H_2, CH_4, CO_2, light and heavy hydrocarbons. These gases can be used for electricity generation. The entire biomass gasification process can be categorized under three sections, viz. upstream processing, gasification, and downstream processing. The biomass is processed before subjected to gasification, where it is shredded, dried, and densified. This is essential because of suitable particle size, low moisture content, and low efficiency. The gasification process takes place at a range of 600°C to 1000°C in the presence of gasifying agents (N_2/CO_2/air/steam). In that case, it helps in converting heavy tar (generated during primary pyrolysis of biomass) into light hydrocarbon, char, ash, and permanent gases (H_2, CO, CO_2, etc.) (Table 1.3). In recent years, catalytic biomass gasification technologies have evolved in leaps and bounds. The use of a catalyst such as Fe and Cr has improved syngas yield significantly. The major bottleneck of this process is catalyst recovery, catalyst poisoning, tar formation, and high ash production.

Environment-friendly and sustainable technologies are the need of the hour. The advent of technologies such as fermentative hydrogen production, biomethane production, biodiesel from oleaginous microorganisms, bioethanol, biobutanol, and bioelectricity generation using biomass-based feedstock has shown potential for large scale clean energy generation.

1.2.1.6.2 Bioethanol

Production of bioethanol has been the oldest of the technologies developed by humans. The use of bioethanol for fuel purposes has gained much attention in recent times. Ethanol has been added to petrol (2 to 10% v/v) to decrease its overall consumption. At present, bioethanol is produced from molasses, starchy feedstock, agricultural waste, and lignocellulosic residues, which makes them genuinely renewable. The metabolic pathways involved in ethanol production involves converting complex polymeric sugars into simple sugars (mainly hexose) through saccharification. These fermentable sugars then undergo anaerobic fermentation where it gets metabolized to produce ethanol (Eq.1.1):

$$C_nH_{2n}O_n \rightarrow \frac{n}{3}CH_3CH_2OH + \frac{n}{3}CO_2 + 227.0 \text{ kJ mol}^{-1} \quad (1.1)$$

The ethanol-rich broth is then subjected to fractional distillation for improvement of its concentration. The ethanol on combustion gives 1370.7 kJ mol^{-1} (Eq. 1.2) of energy that can be harnessed for cooking, automobile combustion engine, etc.

$$CH_3CH_2OH + 3O_2 \xrightarrow{Combustion} 2CO_2 + 2H_2O + 1370.7 \text{ kJ mol}^{-1} \quad (1.2)$$

Some of the salient features of bioethanol are:

- Easy for storage and require minimalistic changes in the existing infrastructure for sustainable distribution.
- It can be easily blended with automobile fuel causing minimum damage to internal combustion engines.
- The emission profile of ethanol is shallow as compared to fossil fuel.

TABLE 1.3
Summarizes All the Required Parameters for the Biomass Gasification Process

Process	Process parameters	Observations
Biomass	1. Biomass types	a) The comparison of biomass viz, cellulose, hemicellulose, and lignin influence the composition of the syngas. b) Higher (hemicellulose + cellulose)/lignin ratio results in higher syngas yield.
	2. Gravimetric moisture content	a) Superior energy efficiency and product yield would be possible if the low moisture content is present in the biomass. b) Greater tar formation will be observed if biomass contains 30–40% w/w moisture content. c) Ideally, moisture content should be in a range of 10–20% w/w. This moisture content helps in maintaining a stable bed temperature.
	3. Ash content	If the ash content of biomass is less than 2% w/w, then such biomass is considered suitable for gasification. The common biomass residues, oilseeds, husks, grasses, have ash content > 10% w/w.
	4. Particle size	a) Higher surface area, lower diffusional resistance can be achieved with smaller particle size. b) Smaller particle sizes (0.15 to 50 mm) are suitable for gasification. c) In a fixed bed reactor, particle size> 51 mm can be processed. d) With the increase in gasification, the temperature can reduce the influence of particle size on the process.
	5. Temperature	a) Woody biomass and agricultural residues gasify at the temperature at 850–950°C and 750–850°C, respectively. b) Hemicellulose generally degrades at the temperature range of 250–300°C, whereas lignin and cellulose above 400°C. c) The higher reduction can be achieved at higher operating temperatures.
Gasification parameters	6. Bed material	a) Bed material could be inert or active. b) During biomass gasification, the energy transferring medium could be used as inert bed material.
	7. Steam to biomass ratio (S/B)	a) The S/B ratio of 0.3–1.0 has shown improved gasification efficiency. b) Higher H_2 yields have been observed with low tar formation at an S/B ratio of 1.35–4.04. c) Higher steam in the gasifier may lead to excessive tar formation since it might decrease the temperature inside the reactor.

(continued)

TABLE 1.3 (continued)
Summarizes All the Required Parameters for the Biomass Gasification Process

Process	Process parameters	Observations
	8. Gasification agents	a) Air, O_2, steam, and CO_2 are commonly used gasification agents. b) When the air was used as a gasification agent, syngas with heating value was between 4 and 7 MJ/Nm³ was produced. c) When steam was used as a gasification agent, a high H_2 yield was observed due to the water gas shift reaction. d) Whereas using O_2 for gasification resulted in the formation of syngas with a high heat value due to higher H_2 concentrations but it is an expensive gasifying agent.
	9. Equivalence ratio	a) An equivalence ratio of 0.2 to 0.3 has been reported optimum for improved gasification efficiency. b) An inverse relationship has been observed between H_2 and CO yield in syngas with the equivalence ratio. c) at an equivalence ratio above 0.4, complete gasification is achieved, whereas it remains incomplete below 0.2.

Source: Ren et al., 2018

A handful of countries have access to the technologies and skills for ethanol production. The developing countries are now looking forward to producing and blending ethanol with automotive fuels to reduce the import burden associated with fossil fuels. The use of energy crops for ethanol production may hold much promise and fueled the debate of "food vs. fuel." Unforeseen anthropogenic challenges such as the COVID-19 pandemic have always led to stress on global food reserves. Thus, tinkering with agricultural practices for fuel generation could have a severe impact on human food security.

The use of lignocellulosic biomass residues, which are otherwise treated as waste, can be a potential feedstock for ethanol production. To tap the sugars trapped in the polymeric form in lignocellulosic biomass, they must be subjected to certain pre-treatment processes (physical, chemical, physicochemical, and biological pre-treatment) followed by saccharification. Lignin is the nonfermentable polymer that needs to be removed so that complex polymers such as cellulose and hemicellulose can be depolymerized.

1.2.1.6.3 Biobutanol

In search of liquid biofuel gained much importance due to their ease in transportation exploiting the existing infrastructure. This has led to an expression of interest towards biobutanol production through anaerobic acetone–butanol–ethanol (ABE) fermentation. The exploitation of *Clostridium* sp., for biobutanol production has shown promise in terms of yield and substrate conversion efficiency. The problem with biobutanol production is low yield leading to non-suitability for commercialization. Moreover, various other non-desired by-products are also produced along with butanol which caused the problem during downstream processing. Biobutanol production consists of two stages viz. acidogenesis followed by solventogenesis. Production of H_2 and exhaustion of NADH are two factors that decrease the biobutanol yield. An overhead pressure is needed to be applied to initiate solventogenesis. Maintenance of strict anaerobic conditions throughout the process is cumbersome and thus makes the overall process expensive.

1.2.1.6.4 Biodiesel

Biodiesel is a mixture of monoalkyl esters derived from different oil-rich resources. The raw material for biodiesel is acyl fatty acids that are sourced from various natural resources, viz. plants origin oils, animal fat, oleaginous microorganisms, etc. The acyl fatty acids are subjected to a transesterification process to convert them into more volatile esters. The ester bond present between glycerol and fatty acid is broken through acid or base catalysis in the presence of methanol. This results in the formation of fatty acid methyl esters (FAME), as shown in Eq. 1.3.

$$\text{Triglyceride} + 3\ \text{Methanol} \xrightleftharpoons{\textit{Catalyst}} \text{Glycerine} + 3\ \text{Methyl Esters}\left(\text{Biodiesel}\right) \quad (1.3)$$

The use of catalyst improves the conversion efficiency toward FAME formation. Homogenous acid catalysts such as HCl or alkaline catalysts such as potassium hydroxide are commonly used in transesterification. The use of heterogeneous catalysts such as CaO, Zeolite Socony Mobil (ZSMF), Fe_2O_3, etc. has shown promising improvement in FAME yield. As the transesterification reaction is reversible in nature, the introduction of methanol in the reaction mixture could shift the equilibrium towards the forward direction. The solvent, i.e., methanol, can be recovered from the process through the Soxhlet technique and rotatory evaporation. The entire transesterification reaction takes place at 60°C. The kinetics of the reaction can be increased towards the forwarding direction by sequential increasing the reaction temperature to 80°C could increase the kinetics of the reaction. The reaction time of 90 min showed maximum reactant conversion.

Biodiesel was initially made from edible derived from plants such as soybean, rapeseed, canola, sunflower, palm, and coconut. It may lead to the "food versus fuel" debate in a larger context as there is already a shortage of edible oil for human consumption. A country like India imports most of its edible oil needs, so biodiesel technologies based on edible oil are not feasible. Among different oleaginous seeds, biodiesel production using *Jatropha curcas* seed was found to be more promising in terms of yield and productivity. Jatropha oil is nonedible, and per hectare, yield is high as compared to other oil seeds. Jatropha oil has low acidity, good stability, low viscosity. Besides this, it has a higher cetane number compared to diesel. This makes it an excellent alternative fuel with no modifications required in the engine.

Recently, it has been widely reported that many oleaginous microorganisms produce oil under stressed conditions. Microorganisms such as algae, yeast, and mold are known for their lipid production ability (Kumar et al., 2016). A distinct shift in lipid profile has been observed when oleaginous microorganisms grew under favorable and stressed conditions. When the conditions are favorable for growth, microorganisms grew rapidly. The lipid produced under this condition is basically for the maintenance of cell membrane integrity, consisting of about 5–20% w/w of their dry cell weight. The lipid profile constituted medium-chain (C10–C14), long-chain (C16–C18), very long-chain (C20), and fatty acid derivatives. The entire lipid profile changes once there is an onset of stressed conditions where the microorganisms channelize their metabolic machinery to produce and accumulate more neutral lipids (20–50% w/w of dry cell weight). Major constituents of such lipids are in the form of triacylglycerol (TAG). The role of TAGs is not performing a structural role in the phospholipids cell membrane; rather, it serves as storage energy (Hu et al., 2008). The major bottleneck of microbe-derived biodiesel is their low yields, cost of operation of the bioreactor, and higher risk of undesired contamination.

1.2.1.6.5 Biohydrogen

Hydrogen can be considered ideal fuel in terms of its high energy density and clean emission profile. It is the only fuel that on burning does not have any carbon footprint. Biological hydrogen (H_2) production takes place mainly at ambient temperatures and pressures, making this process less energy-intensive than other conventional processes (chemical or electrochemical process). Microbial species capable of producing H_2 belong to different taxonomic and physiological types. The pivotal enzyme complex involved in H_2 production is hydrogenase or nitrogenase. These enzymes regulate the hydrogen production process in prokaryotes and some eukaryotic organisms, including green algae. The excess electrons generated during catabolism inside the cells are disposed of in the form of H_2 by the function of hydrogenase protein.

The following processes mainly produce Bio-H_2: a) photolysis of water (direct and indirect biophysics) by blue-green algae and microalgae and b) oxidation of organic acids by photofermentation and dark fermentation (using mesophilic or thermophilic bacteria). However, the processes mentioned above have their advantages and disadvantages.

The H_2 is produced through photolysis of water catalyzed by photosystem complexes of autotrophs and photofermentation of organic acids by photo-organotrophs. However, for both cases, the H_2 production rate (HPR) has been relatively low. Moreover, these microorganisms require specialized photobioreactor, which makes the overall process labor-intensive and commercially nonviable. In terms of HPR and yield, dark fermentative pathways hold promise as it is independent of light energy and requires moderate process conditions (Figure 1.4).

Under anaerobic conditions, the fate of pyruvate is to get converted into acetic acid and butyric acid. In doing so, H_2 is produced as a by-product. The

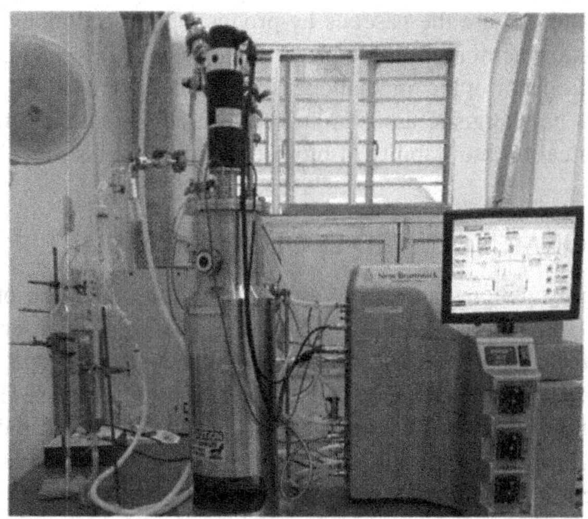

FIGURE 1.4 A 20 L bioreactor for bioH_2 production.

pyruvate-ferredoxin oxidoreductase (PFOR) enzyme oxidizes pyruvate to acetyl coenzyme A (acetyl-CoA). This pyruvate oxidation step requires ferredoxin (Fd) reduction which in its reduced form is oxidized by [FeFe] hydrogenase and catalyzes the formation of H_2. The overall reaction is shown in Eq. 1.4 and Eq. 1.5.

$$\text{Pyruvate} + \text{CoA} + 2\ \text{Fd (ox)} \rightarrow \text{Acetyl-CoA} + 2\ \text{Fd (red)} + CO_2 \tag{1.4}$$

$$2\text{H+} + \text{Fd(red)} \rightarrow H_2 + \text{Fd(ox)} \tag{1.5}$$

Hydrogen production can be considered renewable as the feedstock used for its production are generally lignocellulosic biomass, agriculture wastes, molasses, etc. such a process can augment the existing H_2 production technologies. Moreover, such installations can be installed in remote areas; thus, a decentralized energy infrastructure could be developed in near future.

1.2.1.6.6 Biomethane

Biogas is also known as marsh gas, is rich in methane with impurities of H_2S, CO_2, and H_2 was first time reported by Volta. Commonly used feedstock for anaerobic digestion is organic residues, sludge, manure, agriculture residues, and lignocellulosic biomass. The biomethanation is a complex process that involves interplay between various microbial metabolic pathways. The complex biopolymers such as carbohydrates, protein, and fats are hydrolyzed into monomeric constituents by the action of hydrolytic enzymes secreted by heterotrophic microorganisms. These monomers were further channelized for the acidogenic pathway under anaerobic conditions, where they are subsequently metabolized into short-chain fatty acids (SCFAs) such as acetic acid, propionic acid, butyric acids and ethanol.

The H_2, CO_2, and CO are the gaseous by-products produced along with SCFAs. Then acetoclastic microorganisms convert these SCFAs into methane and CO_2. Moreover, hydrogenotrophic methanogens consume overhead H_2 and CO_2 to methane. Thus, a synergistic association between acidogenic and methanogenic microbes is critical for the stability of biomethanation process.

1.3 ENVIRONMENTAL AND ECOLOGICAL IMPACT

If biofuel-friendly policies are to be implemented, they must be promoted as a panacea to mitigate global warming, restore forest cover, and conserve ecological hemostasis. Much emphasis is needed to implement technologies that would improve the biofuel efficiency generated through biological routes. The production of first-generation biofuels utilizing available feedstocks would reduce 20–60% in emissions compared to fossil fuels (FAO, 2008). Implementation of green biofuel production technologies will prove beneficial to nature soon and secure energy requirements for future generations. Biomass-based biofuel and bioenergy technologies will promote the upliftment of lower-income society associated with the maintenance, processing, and collection of such resources. Rotation of conventional crops with energy crops can also increase the income of farmers and eventually helps in soil conservation. If biomass is exploited scientifically, it can help in the sequestration of atmospheric carbon dioxide, thereby reducing the overall footprint of the energy generation process. The development of an efficient distributed energy generation system can integrate various green energy technologies; thus, better utilization of resources can be achieved.

REFERENCES

Assadi, M. Khalaji, S. Bakhoda, R. Saidur, and H. Hanaei. "Recent progress in perovskite solar cells." *Renewable and Sustainable Energy Reviews* 81 (2018): 2812–2822.

Barber, Daniel. "Public perceptions of tidal energy between Australia and Canada." (2020). Bachelor's thesis, Novia University of Applied Sciences, Finland.

Bergman, Patrick CA, and Jacob HA Kiel. "Torrefaction for biomass upgrading." In *Proc. 14th European Biomass Conference, Paris, France* (2005), pp. 17–21.

Bergman, Patrick CA. "Combined torrefaction and pelletisation: the TOP process." Netherlands (2005), Available at www.ecn.nl/docs/library/report/2005/c05073.pdf or via www.ecn.nl/library/reports/2005/c05073.html from the Energy research Centre of the Netherlands (http://www.ecn.nl/), Postbus 1, 1755 ZG Petten (NL).

Black, Brian C. *Crude Reality: Petroleum in World History*. Rowman & Littlefield (2014).

BP Energy outlook (2019), www.bp.com/content/dam/bp/business-sites/en/global/corporate/pdfs/energy-economics/energy-outlook/bp-energy-outlook-2019.pdf

Brock, William A., Gustav Engstrom, and Anastasios Xepapadeas. "Energy balance climate models, damage reservoirs and the time profile of climate change policy" (2012). FEEM Working Paper No. 20.2012, Available at SSRN: https://ssrn.com/abstract=2040674 or http://dx.doi.org/10.2139/ssrn.2040674

Cho, Y.-S., J. W. Lee, and W. Jeong. "The construction of a tidal power plant at Sihwa Lake, Korea." *Energy Sources, Part A: Recovery, Utilization, and Environmental Effects* 34, no. 14 (2012): 1280–1287.

Dagem D, Gebremichael, and Youngsoo Jung. "Reactor Types and Physical Breakdown Structure (PBS) for Nuclear Power Plant." Proceedings of the 2020 Spring Architectural Institute of Korea (AIK), 40, no. 1 (2020): 432–433.

Das, Debabrata, and Shantonu Roy. *Biohythane: Fuel for the Future*. CRC Press (2016).

de Oliveira Azevêdo, Rômulo, Paulo Rotela Junior, Gianfranco Chicco, Giancarlo Aquila, Luiz Célio Souza Rocha, and Rogério Santana Peruchi. "Identification and analysis of impact factors on the economic feasibility of wind energy investments." *International Journal of Energy Research* 45, no. 3 (2020): 1–27.

Dearne, Martin J., and Keith Branigan. "The use of coal in Roman Britain." *The Antiquaries Journal* 75 (1995): 71–105.

Energy Technology Prospective 2020, IEA, Global primary energy demand by fuel, 1925–2019, IEA, Paris, www.iea.org/data-and-statistics/charts/global-primary-energy-demand-by-fuel-1925-2019

Food & Agriculture Organization, Agriculture Organization. *The State of Food and Agriculture 2008: Biofuels: prospects, risks and opportunities*. Vol. 38. 2008.

Hu, Qiang, Milton Sommerfeld, Eric Jarvis, Maria Ghirardi, Matthew Posewitz, Michael Seibert, and Al Darzins. "Microalgal triacylglycerols as feedstocks for biofuel production: perspectives and advances." *The Plant Journal* 54, no. 4 (2008): 621–639.

Kim, Kun Joong, Moran Balaish, Masaki Wadaguchi, Lingping Kong, and Jennifer LM Rupp. "Solid-State Li–Metal Batteries: Challenges and Horizons of Oxide and Sulfide Solid Electrolytes and Their Interfaces." *Advanced Energy Materials* (2020): 2002689.

Kumar, Kanhaiya, Supratim Ghosh, Irini Angelidaki, Susan L. Holdt, Dimitar B. Karakashev, Merlin Alvarado Morales, and Debabrata Das. "Recent developments on biofuels production from microalgae and macroalgae." *Renewable and Sustainable Energy Reviews* 65 (2016): 235–249.

Lu X., Yamauchi K., Phaiboonsilpa N., Two-step hydrolysis of Japanese beech as treated by semi-flow hot-compressed water, *Journal of Wood Science*, 55 (2009): 367–375.

Morison, Rachel. "U.K. Wind Drought Heads Into 9th Day With No Relief for Weeks" Bloombergquint.com, 7th June 2018, www.bloombergquint.com/technology/u-k-wind-drought-heads-into-9th-day-with-no-relief-for-weeks

Nayak, A. K., Arun Kumar, P. S. Dhami, C. K. Asnani, and P. Singh. "Thorium Technology Development in an Indian Perspective." In *Thorium—Energy for the Future*, pp. 27–82. Springer, Singapore (2019).

Nayak, Bikram K., Shantonu Roy, and Debabrata Das. "Biohydrogen production from algal biomass (Anabaena sp. PCC 7120) cultivated in airlift photobioreactor." *International journal of hydrogen energy* 39, no. 14 (2014): 7553–7560.

ONE, Advantages and Challenges of Nuclear Energy (2021) Retrieved from: www.energy.gov/ne/articles/advantages-and-challenges-nuclearenergy.

Perlack, R.D., Wright, L.L., Turhollow, A.F., Graham, R.L., Stokes B.J., and Erbach, D.C. 2005. Biomass as Feedstock for a Bioenergy and Bioproducts Industry: The Technical Feasibility of a Billion-Ton Annual Supply. Tech. Rep. DOE/GO-102005-2135 and ORNL/TM-2005/66, April.

Popp, József, Sándor Kovács, Judit Oláh, Zoltán Divéki, and Ervin Balázs. "Bioeconomy: Biomass and biomass-based energy supply and demand." *New Biotechnology* (2020).

Prieto, Cristina, Rafael Osuna, A. Inés Fernández, and Luisa F. Cabeza. "Thermal storage in a MW scale. Molten salt solar thermal pilot facility: Plant description and commissioning experiences." *Renewable Energy* 99 (2016): 852–866.

Quintana-Rojo, Consolación, Fernando-Evaristo Callejas-Albiñana, Miguel-Ángel Tarancón, and Pablo del Río. "Assessing the feasibility of deployment policies in wind energy systems. A sensitivity analysis on a multiequational econometric framework." *Energy Economics* 86 (2020): 104688.

Ren, Jie, Jing-Pei Cao, Xiao-Yan Zhao, Fei-Long Yang, and Xian-Yong Wei. "Recent advances in syngas production from biomass catalytic gasification: A critical review on reactors, catalysts, catalytic mechanisms and mathematical models." *Renewable and Sustainable Energy Reviews* 116 (2019): 109426.

Tester, Jefferson W., Brian J. Anderson, A. S. Batchelor, D. D. Blackwell, Ronald DiPippo, E. M. Drake, J. Garnish et al. "The future of geothermal energy." Massachusetts Institute of Technology 358 (2006): 1–13.

Thurman, Harold V., and Elizabeth A. Burton. *Introductory Oceanography*. New York: Prentice Hall (1997).

Zhang, Yixiang, Zongxi Zhang, Yuguang Zhou, and Renjie Dong. "The Influences of Various Testing Conditions on the Evaluation of Household Biomass Pellet Fuel Combustion." *Energies* 11, no. 5 (2018): 1131.

2 Dark Fermentative Hydrogen Production

2.1 INTRODUCTION

The world's population and energy needs are growing at a faster rate, resulting in a significant rise in the use of fossil fuels. The continuing fossil resources depletion as well as their rising prices, is widening the gap between the industrialized world's energy needs. The emission of GHGs ("Greenhouse Gases") and pollutants increases as a result of the burning of fossil fuels. The unrestricted use of fossil fuels has negative environmental and ecological consequences, prompting a renewed interest in creating renewable and sustainable energy sources to minimize dependency on fossil fuels. There is widespread concern about global warming and the depletion of energy resources. As a result, there is a strong need to find renewable energy alternatives to traditional fossil fuels that are safe, effective, economical, sustainable, and eco-friendly while still meeting rising energy requirements. Renewable energy generation is a critical and difficult undertaking for achieving a cleaner environment, reducing the price volatility of fossil fuels, and ensuring energy security. Biofuels might be taken as an alternative, eco-friendly, and sustainable energy source for fossil fuels in order to maintain a clean environment, save forgiven currency, and prevent pollution.

Hydrogen is seen as a long-term, clean energy carrier with high energy production (122kJ g^{-1}), and hence has the potential to become the primary fuel source of the future (Roy and Das, 2015). Hydrogen has a high calorific value of approximately 3042 cal m^{-3} and the greatest gravimetric energy density, making it a popular transportation fuel and source of power. Biohydrogen is currently produced primarily through natural gas steam reforming and water electrolysis. On the other hand, these traditional techniques, are not deemed renewable energy resources as they need substantial amounts of energy, which is obtained from burning fossil fuels (Diam et al., 2009).

2.2 MICROBIAL PERSPECTIVE ON DARK FERMENTATIVE HYDROGEN-PRODUCING MICROORGANISMS

Microorganisms have been given a natural option for converting biomass to hydrogen for ages. Production of hydrogen is possible by photosynthetic bacteria,

DOI: 10.1201/9781003224587-2

methylotrophs, aerobes, facultative anaerobes, and anaerobes (Sengupta and Nandi, 1998). Though, in the biosphere, biologically generated hydrogen is mostly produced by microbial fermentation activities.

2.2.1 Mesophilic Dark Fermentative Hydrogen Production

Under anaerobic circumstances, a vast range of bacteria creates hydrogen in the natural world. Different bacteria have been identified in various parts of the globe, each with a distinct capacity to produce hydrogen. The hydrogen generation was recorded in several microbe areas, including photosynthetic bacteria, methylotrophs, facultative anaerobes, and anaerobes (Nandi and Sengupta, 1998).

Strict and facultative anaerobic chemoheterotrophs related to *Enterobacteriaceae* and *Clostridia* are among the most promising hydrogen-generating bacteria (Figure 2.1). Microorganisms produce hydrogen to discharge excess electrons produced during substrate oxidation inside the cells.

2.2.1.1 Hydrogen-producing *Clostridium* sp.

These obligate anaerobes are fermentative and produce spores. They are gram-positive, rod-shaped, microorganisms with a low concentration of G + C. It has a shorter doubling time and could resist adverse circumstances (physical stress, heat shock, and so on) when compared with other anaerobic microorganisms. The attributes listed above might be deemed industrially significant. The fermentative use of the *Clostridia* strain for solvents as well as alcohol formation was considered during World War I (Weizmann et al., 1937). Hydrogen is one of the byproducts of such a solventogenic process. This group of species has been found to be among the highest hydrogen yields (Kamalaskar et al., 2010). Aa few recently isolated "*Clostridium*" sp. like C. *pasteurianum, C. beijerinckii, C. welchii*, and *C. butyricum* were utilized independently or as synthetic blended consortia for hydrogen production. The termite gut-isolated *C. beijerinckii* AM21B had the maximum hydrogen output, ranging from 1.8–2.0 mol mol^{-1} glucose (Taguchi et al.,1996). Other carbohydrates that this strain may use include fructose, sucrose, cellobiose, galactose, arabinose, and xylose. *Clostridium sp.* has also been shown to utilize hemicellulose and cellulose which are prevalent in lignocellulosic biomass, as a substrate for hydrogen generation.

2.2.1.2 Hydrogen-producing *Enterobacter* sp.

These microorganisms are mainly facultative anaerobes, rod-shaped, gram-negative, motile, or non-motile. In comparison to obligate anaerobes, they have faster growth rates. They may use a variety of carbon sources and are dissolved oxygen resistant at low levels. The hydrogen production was not hindered by the presence of high pH_2 ("Partial Pressure of Hydrogen"). However, when employing glucose as a substrate, *Enterobacter sp.* produces less hydrogen than *Clostridium sp.* In batch fermentation, yield and rate of hydrogen generation of 1.0molmol^{-1} glucose and 21mmol $L^{-1}h^{-1}$ were recorded (Tanisho, 1994).

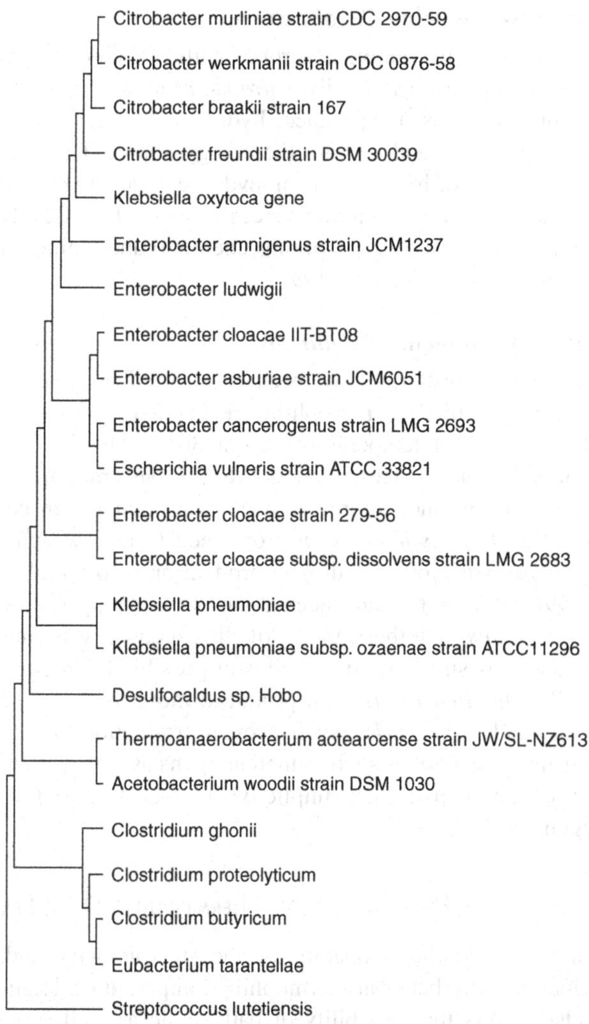

FIGURE 2.1 Phylogenetic tree of various hydrogen-producing microbial species. The *Streptococcus lutetiensis* has been used as outgroup.

2.2.1.3 Hydrogen-producing *Escherichia* sp.

These are usually rod-shaped, gram-negative, motile microorganisms, with low concentrations of G+C. Mainly, it makes hydrogen from formate. Formate is transformed to CO_2 and hydrogen in the lack of oxygen catalyzed with enzyme complex FHL ("Formate Hydrogenlyase") (Stickland, 1929). The formate dehydrogenase and hydrogenase subunits of FHL are membrane-bound multi-enzyme complexes. The hydrogen output was 0.9 to 1.5 mol mol^{-1} glucose when *E. coli* was used (Rosales-Colunga et al., 2015).

2.2.1.4 Hydrogen-producing *Citrobacter* sp.

These are gram-negative, facultative anaerobes with low G+C concentration and belong to the *Enterobacteriaceae* family. *Citrobacter* uses both chemolihotrophic and organotrophic methods to produce hydrogen. Under chemolihotrophic conditions where hydrogen and CO are the energy source. *Citrobacter sp.* Y19 generated 15 mmol $L^{-1}h^{-1}$ of hydrogen using hydrogen and CO through the water-gas shift process under anaerobic circumstances (Jung et al., 2002). The hydrogen output was 1.1 mol mol^{-1} of glucose during chemoorganotrophic circumstances with glucose as the substrate (Vatsala, 1992).

2.2.1.5 Hydrogen-producing *Bacillus* sp.

Several *Bacillus* genus bacteria have been discovered as potential hydrogen producers. They are facultative mesophilic gram-positive bacteria in general. *Bacillus sp.* thrives best at temperatures about 30°C. They can, nevertheless, withstand considerably greater temperatures. At 37°C, several important enzymes were released. It may produce spores in poor environments. In comparison to *Clostridium* sp., the *Bacillus licheniformis* obtained from cow dung was shown to generate hydrogen with lower hydrogen production of 0.5 mol mol^{-1} glucose (Kalia et al., 1994). The lactic acid mechanism is frequently followed,whereas, the *Bacillus coagulans*, another group of this genus, was obtained from sewage sludge and was similarly identified as a possible hydrogen producer. In comparison to *Bacillus licheniformis*, it produced more hydrogen (2.2 mol mol^{-1} glucose) (Kotay and Das, 2007). For hydrogen production, these strains have benefits over stringent anaerobes such as methanogens as well as *Clostridium* that are industrially essential, like their simplicity of processing and nontoxicity to dissolved oxygen.

2.2.2 Production of Hydrogen from Thermophilic Dark Fermentative

In comparison to mesophilic temperatures, the stoichiometry and kinetics of hydrogen production were better at thermophilic temperatures. High-temperature fermentation also lowers the possibility of pathogenic as well as methanogenic contamination. Contaminations of pathogenic and methanogenic may be passed down from the feedstock or the inoculum, resulting in a reduction in hydrogen production. The dissolved hydrogen in the fermented broth has less of an impact on the hydrogen production in thermophilic regimes. Many industrial sectors release high-temperature, high-organic-content effluents. Food processing effluents, sugar industry wastewater, distillery effluents, and so on are examples of high-temperature industrial effluents. A significant environmental risk might be caused if these effluents are discharged into waterways. An ineffective and sometimes costly method of reducing the biological activity of materials is cooling (Jo et al., 2008). Thermophilic microbes may be able to exploit these high-temperature effluents to produce hydrogen. Thermophilic bacteria may be categorized based on optimal development temperature:

- True thermophiles, which grow at temperatures ranging from 55°C to 75°C,
- mild thermophiles that grow at temperatures between 45°C and 55°C,
- extremophiles are organisms that grow at temperatures exceeding 75°C.

Several thermophilic hydrogen generating species were found like *Caldicellulosiruptor* sp., *Thermoanaerobacter*, *Thermoanaerobacterium*, *Clostridia*, and *Thermotoga* (Figure 2.2) (Zeidan et al., 2010).

2.2.2.1 Hydrogen-producing *Thermoanaerobacterium* sp.

Clostridium species have interconnections with this genus. In 1993, it was discovered in Yellowstone National Park's Frying Pan Springs. It is capable of degrading xylan and producing hydrogen (Lee et al., 1993). These bacteria have gram-negative straight rods and motile peritrichous flagella and comprise lower G+C. They produce spores when starved of nutrients. They also produce lactate, hydrogen, CO_2, acetate, and ethanol among other metabolic end products.

2.2.2.2 Hydrogen-producing *Thermoanaerobacter* sp.

According to *Bergey's Manual of Systematic Bacteriology*, this species belongs to the group of gram-positive rods that seem to be irregular and non-spore-forming (Wiegel and Ljungdahl, 1981). Lactate, ethanol, and hydrogen are fermentation byproducts produced by thermophilic anaerobic bacteria of the genera *Thermoanaerobacter* and *Thermoanaerobium*; non-spore-forming bacteria are obligate anaerobes. Microorganisms may use a wide range of sugars, but cellulose cannot be metabolized by any of them. As a result, the most important byproducts are CO_2, hydrogen, ethanol, lactate, and acetate. These species were not found to produce any butyrate. Hydrogen production may reach up to 4 moles per mole of glucose, which is theoretically the maximum value.

2.2.2.3 Hydrogen-producing *Clostridium* sp.

Recent years have seen a significant increase in the relevance of this genus in the biofuels field. The phylum Firmicutes includes the genus *Clostridium*. They are obligate anaerobic organisms that are gram-positive, rod-shaped, motile, frequently spore-forming, and obligatory anaerobic. They may use cellulase to break down ferment lignocellulosic and cellulose material to produce hydrogen. The greatest hydrogen output of 1.6mol per mol hexose has been reported by cellulose and microbe as a carbon source (Levin et al., 2006). These bacteria have a significant role in the conversion of "lignocellulosic" biomass into biohydrogen as they have enzymatic repertoire for degrading cellulose.

2.2.2.4 Hydrogen-producing *Caldicellulosiruptor* sp.

Extremophile "*Caldicellulosiruptor saccharolyticus*" produces hydrogen and grows at 70°C. This microbe was classified into the *Bacillus/Clostridium* subphylum based on physiological features and phylogenetic location (Rainey et al., 1994). These bacteria must be anaerobic and gram-positive microorganisms that do not

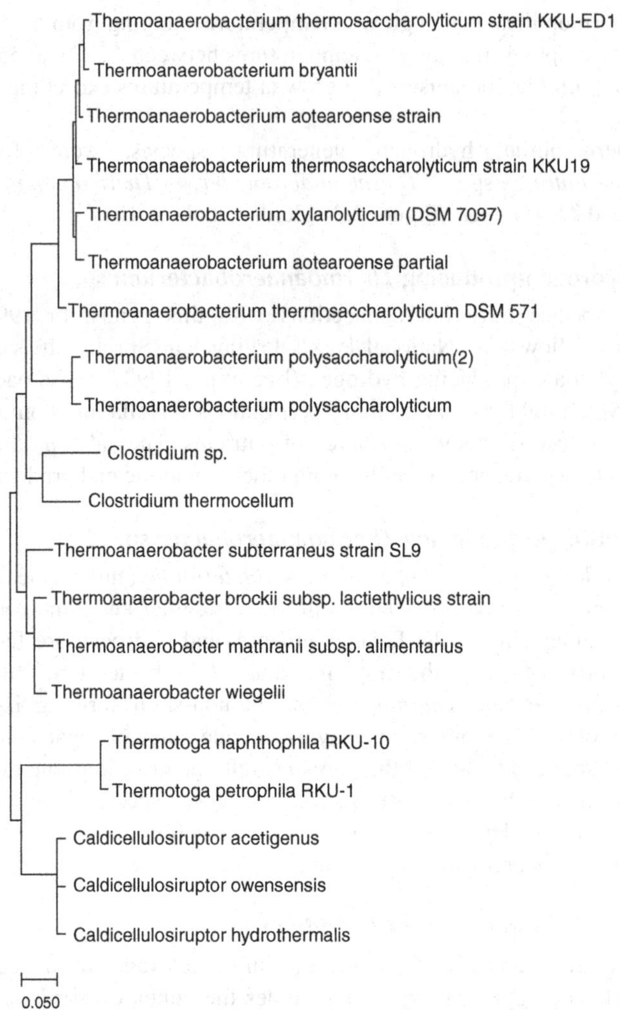

FIGURE 2.2 Phylogenetic tree of various hydrogen-producing thermophilic microbial species.

generate spores. These bacteria have been obtained from natural environments such as lake sediments and hot springs. With the aid of a large variety of hydrolytic enzymes, these bacteria may use a wide variety of substrates like xylose, xylan, cellobiose, as well as cellulose. Hydrogen generation from lignocellulosic waste might benefit from the features of these species. Lactate and acetate are the most common metabolites produced by this microorganism. Surprisingly, in comparison to other thermophilic microorganisms like *Thermoanaerobacterium sp.*, butyrate is not formed as a metabolic end product. Using paper pulp, a maximum volumetric hydrogen generation rate of 5 to 6 mmol $L^{-1}h^{1}$ was observed (Kádár et al., 2004).

2.2.2.5 Hydrogen-producing *Thermotoga* sp.

These microorganisms can grow at 90°C, which is the greatest temperature for hydrogen production ever recorded. It was originally separated from sea bottoms heated by geothermal energy in Italy and the Azores. Their genus name comes from the existence of a distinctive outer sheet-like structure known as a toga. They are gram-positive, rod-shaped obligate anaerobes. Temperatures, pressures, and sulfur levels are all extreme in their native habitats. They may get their electrons from elemental sulfur, thiosulphate, or both. Acetate, hydrogen, and CO_2 are the primary components of their metabolic end products, with ethanol present in trace levels. The *Thermotoga neoplanita* and *Thermotoga maritima* are two microorganisms that have been found to produce hydrogen. On feeding waste glycerol, *Thermotoga* sp. produced 1.98 ± 0.1 mol hydrogen mol^{-1} $glycerol_{consumed}$ (Ngo et al., 2011).

2.3 BIOCHEMISTRY OF DARK FERMENTATIVE HYDROGEN PRODUCTION

Glucose is the primary substrate for the dark fermentative generation of hydrogen. Hydrolysis of complicated polymeric organic material produces simple sugars including glucose. Glucose is the most basic sugar, and it is favored by the majority of bacteria. It is then converted to pyruvate through the glycolytic process. In this way, microbes manufacture their energy source, ATP. Pyruvate is then transformed to butyric and acetic acid in anaerobic conditions. As a result, hydrogen is created as a byproduct (Figure 2.3).

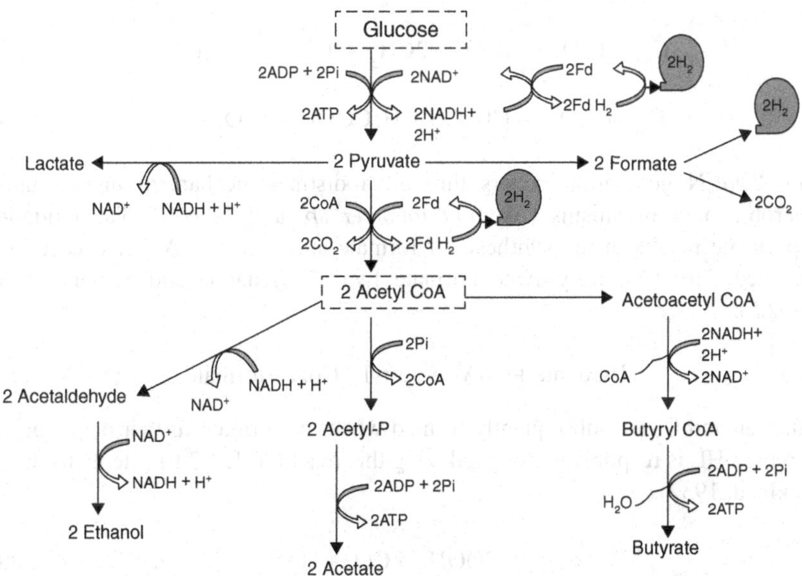

FIGURE 2.3 Metabolic pathways involved in dark fermentative hydrogen production.

During anaerobic growth, cells must deal with the extra electrons created by substrate metabolism (glucose). Therefore, numerous types of particular controls are necessary to govern the flow of electrons in fermentative bacteria's metabolic process. Thus, certain microorganisms evolved the capacity to dispose of extra electrons as molecular hydrogen. Such processes are catalyzed by the hydrogenase enzyme. The hydrogenase enzyme complex is divided into two categories. One of these is uptake hydrogenase, which converts hydrogen into protons and electrons. Hydrogenase is the second enzyme, which lowers protons and releases molecular hydrogen as a result. Pyruvate is a significant source of reducing equivalents for the generation of hydrogen.

For hydrogen production, obligate anaerobes like *Clostridia* and thermophilic bacteria need a unique process. Pyruvate is transformed to acetyl-CoA ("Acetyl Coenzyme A") with enzyme PFOR ("Pyruvate-Ferredoxin Oxidoreductase"). This pyruvate oxidation phase necessitates Fd (ferredoxin) reduction, which is oxidized with Fe-Fe hydrogenase and catalysis the generation of hydrogen in its reduced form. The reaction is revealed in Eq. 2.1 and Eq. 2.2.

$$\text{"Pyruvate + CoA + 2Fd(ox)} \rightarrow \text{Acetyl-CoA + 2 Fd(red) + CO}_2\text{"} \quad (2.1)$$

$$\text{"2H}^+ \text{ + Fd(red)} \rightarrow \text{H}_2 \text{ + Fd(ox)"} \quad (2.2)$$

If acetate is the only result of pyruvate oxidation, stoichiometry reveals that four moles of hydrogen per mole of glucose may be created. A mole of glucose may provide two moles of hydrogen if butyrate is the single end product. Eq. 2.3 and Eq. 2.4 demonstrate the metabolic fate of glucose in terms of butyric and acetic acids along with hydrogen production.

$$C_6 + H_{12}O_6 + 2H_2O \rightarrow 2CH_3COOH + 2CO_2 + 4H_2 \quad (2.3)$$

$$C_6 + H_{12}O_6 \rightarrow CH_3CH_2CH_2COOH + 2CO_2 + 2H_2 \quad (2.4)$$

HYDROGEN generation occurs through a distinct mechanism in facultative anaerobic microorganisms (like *Enterobacter sp.* and *E. coli*). The oxidation of pyruvate results in the synthesis of formate and acetyl-CoA. This reaction is catalyzed with PFL ("Pyruvate Formate Lyase") (Knappe and Sawers, 1990) (Eq. 2.5).

$$\text{Pyruvate} + \text{CoA} \rightarrow \text{acetyl} - \text{CoA} + \text{formate} \quad (2.5)$$

hydrogen and C_2 are subsequently formed when the formate further decomposes. Enzyme FHL is responsible for catalyzing this reaction (Eq. 2.6) (Stephenson and Stickland, 1932).

$$HCOOH \rightarrow CO_2 + H_2 \quad (2.6)$$

2.3.1 Genetic Alteration of a Metabolic Process for Enhancement of Hydrogen Production

The progress of molecular biology and metabolic engineering has ushered in a new era of biofuel research. Microorganisms' metabolic pathways have been modified using methods including gene silencing/deletion, synthetic biology, and homologous overexpression to increase hydrogen production. Implementation of the above-mentioned approaches to divert or redirect electron flow towards hydrogen production pathways could help to achieve near to theoretical yield. Furthermore, increasing substrate utilization efficiency and employing "protein engineering" to create an effective and/or oxygen tolerant hydrogenase might be beneficial in the development of robust strains.

The [FeFe] hydrogenase is encoded by the hydA gene in obligate anaerobes. This enzyme aids in the electron transport from ferredoxin to proton. One of the early studies, heterologous or homologous overexpression of the hydA gene led to improvement in hydrogen production. In comparison to the wild strain, homologous hydA gene overexpression of *Clostridium paraputrificum* M-21 resulted in a 1.7-fold increase in hydrogen generation. The lactic acid mechanism was inhibited in this research, but acetic acid generation was boosted (Morimoto et al., 2005). One of the biggest obstacles faced during dark fermentative hydrogen generation is the buildup of end metabolites that hinder hydrogen synthesis. During fermentative hydrogen generation, side metabolic processes compete for NADH, lowering the NADH pool available for hydrogen production through the NFOR process, resulting in lower production of hydrogen. Another way is to stop existing side processes from diverting NADH away from hydrogenase enzymes.

Clostridium sp. mutants were created using a site-directed mutagenesis method. In this kind of mutagenesis, homologous recombination is critical. The ClosTron system uses a bacterial group 2nd intron to aid in the generation of stable *Clostridium* mutants. The *Lactococcus lactis* (Ll. ltrB) mobile group 2nd introns spread towards a particular target location through an RNA-mediated "retrohoming" process. It is determined that retrohoming has been successful when the target site DNA and excised intron lariat RNA have complementary base pairings (Mohr et al., 2000). The sequence of DNA encoding the relevant component of an intron may be changed to vary the intron's specificity toward the target DNA. ClosTron-mediated mutagenesis was utilized to create mutants with impaired lactate dehydrogenase activity in *Clostridium pefringens* strain W11. The hydrogen production increased by 51% in the *ldh* deleted mutant strain (Wang et al., 2011). In recent years, knocking down genes whose products compete with the hydrogen generation pathway has become more important. The pfl gene, which encodes PFL, is knocked out, causing the pyruvate pool to be diverted towards the synthesis of formate. Likewise, knocking down the adhE gene prevents acetyl-CoA from being converted to ethanol. A novel way to improve hydrogen production has been developed using a gene deletion procedure that involves a double loss in hydrogenase I and II or disrupting proteins involved in absorption hydrogenase development. Hydrogen may be produced by deleting the gene accountable for the Tat ("Twin-Arginine Translocation") mechanism in *E. coli* MC4100. By deleting the Tat system, it is possible to impede the proper assembly of formate dehydrogenases along with hydrogenases (Fdh-N and Fdh-O) (Redwood et al., 2008).

For hydrogen production, the approach of random mutagenesis utilizing chemical mutagens was also used. Employing proton suicide techniques, random mutants with a deficiency in the alcohol synthesis pathway generated less volatile fatty acid. NADH is required for the development of different metabolites including butanediol, ethanol, butyric, and lactic acid. The metabolic processes that compete for the pool of NADH are inhibited in these mutants due to a fortuitous mutation. In *Enterobacter aerogenes*, improved hydrogen production was demonstrated after random chemical mutagenesis (Lu et al., 2011). Batch fermentation resulted in a 42% increase in hydrogen output overall. In *K. pneumoniae* IIT-BT 08, the random mutagenesis approach was also used to increase hydrogen generation (previously recognized as *Enterobacter cloacae* IIT-BT 08). Blocking alcohol production processes and certain organic acids were used to shift metabolic flux towards accumulation higher NADH pool. Higher NADH concentrations could result in increased hydrogen production. The mutants with a faulty ethanol pathway and lower end acid levels were screened using proton suicide method. In proton suicide screening, the mutants and wild types are exposed to allyl alcohol (7mM) and $NaBrO_3$ and NaBr (40mM each at pH 5.5). Under such conditions, wild type will not survive but mutants with faulty ethanol pathway would survive. Amongst the different mutants, very few mutants were found to have greater hydrogen production than the wild-type strain. Mutagenesis of the "*K. pneumoniae* IIT-BT08" strain resulted in a 1.5 fold increase in the production of hydrogen (Kumar et al., 2001). Several plasmids were found in the *K. pneumoniae* IIT-BT08 strain. Plasmid maintenance needs the expenditure of energy. When the "*K. pneumoniae* IIT-BT08" plasmids were cured, hydrogen production exhibited modest improvement (Khanna et al., 2012).

2.4 CONCEPT OF CONSORTIA DEVELOPMENT

Simple sugars/soluble fermentable sugars have been used in the majority of dark fermentative hydrogen production. The development of a mixed microbial consortium that would harbor various symbiotically related distinct set of microorganisms which could metabolize complex substrates could eventually help in realizing "waste to energy" concept. It is possible that a single strain of bacterium might lacks all of the hydrolytic enzymes needed to break down complex organic molecules such as cellulose, whereas a cocktail of hydrolytic enzymes would be produced by mixed microbial consortia which can be exploited for solubilization of complex carbohydrates found in organic waste. The soluble fermentable sugars might then be used to generate hydrogen. The natural microbial flora contains various kinds of microorganisms like acetoclastic electrogene, methanogenic microorganisms, hydrogen consuming bacteria and hydrogen generating bacteria. Enrichment is the process that promotes hydrogen-producing microorganisms present within a mixed microbial community.

During the process of enrichment, artificial selection pressure was used to encourage hydrogen-producing bacteria while excluding non-hydrogen producers. For enrichment procedures, several pretreatment techniques were investigated. Various techniques such as chemical (organic, alkali, acid), and physical (ultraviolet, ultrasonic, heat, thawing/freezing) were frequently utilized pretreatment approaches (Fang and Liu, 2002). Most of the hydrogen-consuming microorganisms are non-spore formers. The heat shock pretreatment helps in elimination of the non-spore formers. Moreover, methanogens belong to non-spore-forming archaebacteria which also get eliminated with heat shock treatment. Adverse conditions such as high temperature, acidity, alkalinity promotes endospore formation in some groups of bacteria which also encompass hydrogen-producing bacteria. Thus, when favorable conditions were provided, the endospores germinate and the hydrogen forming bacteria dominates the system. Chemical pretreatment such as Bromo Ethane Sulfonate (BES) treatment also helps inhibit methanogens. The BES acts as a competitive inhibitor of Coenzyme M which is responsible for methane production. The existence of symbiotically related microorganisms is among one of the many benefits of developing a functioning microbial consortium. A symbiotic consortium can provide a pool of hydrolytic enzymes that are not found in the hydrogen producer microorganisms.

With the use of "metagenomics" techniques, it is possible to identify the microbes associated with an enriched mixed consortium (Tolvanen and Karp, 2011). Using this new genomics approach, researchers may be able to identify potential hydrogen producers in the enriched mixed consortium. Methods like ribotyping accompanied by DGGE ("Denaturation Gradient Gel Electrophoresis") might aid in extracting the information about the mixed culture's microbial profile. Therefore, employing complex organic wastes (starchy materials, complex cellulosic, lipids, and proteins) for maximal energy extraction might be made easier with the use of appropriate enriched cultures.

2.5 PROCESS PARAMETERS INFLUENCING DARK FERMENTATIVE HYDROGEN PRODUCTION

2.5.1 The Role of pH in Dark Fermentation

The pH is among the most critical chemical parameters in every biological process. It not only controls the effectiveness of microorganisms' enzymatic machinery but also affects the cells' oxidation-reduction potential. Several enzymes are involved in the metabolic process that leads to hydrogen production. The glycolytic enzymes as well as supporting enzymes (formate lyase, Fe-Fe Hydrogenase, and so on) play a crucial role in its production. Pyruvate is formed when glucose is metabolized via glycolysis. The hydrogen yield is concomitant with the metabolic fate of pyruvate under anaerobic condition (Antonopoulou et al., 2008). Because all enzymes have an ideal pH for their optimum activity, it is critical to investigate the impact of pH in hydrogen production. As dark fermentation occurs, the pH profile shifts. The metabolites accumulation such as volatile fatty acids causes a pH decrease. The enzymatic machinery included in hydrogen generation may be affected by the pH lowering. Therefore, extremely low pH (3.8 to 4.2) leads to a stop in the production of hydrogen. Accumulated volatile fatty acids might compromise the integrity of the cell membrane, causing internal pH to be disrupted. Furthermore, at low pH, a metabolic change from acidogenesis to solventogenesis has been seen (Khanal et al., 2004). The inoculum for the hydrogen production process is enriched blended anaerobic or consortia sludge; a lower pH inhibits methanogenesis and other hydrogen consuming microorganisms (Ginkel and Sung, 2001). Many studies have looked at hydrogen production under controlled pH because of the relevance of pH. Under regulated pH conditions, hydrogen generation and substrate conversion improved. There were several reports on hydrogen generation utilizing pentose sugars. The controlled pH of 5.5 to 6.5 indicated substantial enhancement in hydrogen output with maximum hydrogen output of 1.72 mol mol^{-1} of xylose at a pH of 6.5 (Calli et al., 2008). Under a controlled pH of 6.5, the hydrogen production by "*Enterobacter cloacae* IIT-BT 08" increased by 31%, while the efficacy of substrate conversion was increased by 10%.

2.5.2 The Role of Temperature in Hydrogen Production

In the production of hydrogen, the temperature is also important parameter. Temperature regulates metabolism by modulating enzymatic processes, similar to how pH does. Each enzyme has a maximum productivity at specific temperature ranges. Denaturation of metabolic and life-sustaining enzymes would occur at high temperatures. The hydrogen output of 1.7 mol mol^{-1} glucose was reported when the temperature was increased from 15 to 35°C. (Lee et al., 2006). The temperature of the operating environment has a major impact on the metabolic processes and microbial community. Thus, various research had been conducted to better understand the impact of temperature variation on microbial community dynamics (Sinha and Pandey, 2011). When compared to their mesophilic counterparts, thermophilic hydrogen-producing bacteria produced more hydrogen (Van Groenestijn et al., 2002). Thermodynamically, higher temperatures make the hydrogen generation process more favorable. It raises the system's entropy, increasing total free energy and making the process spontaneous.

2.5.3 The Role of Partial Pressure in Hydrogen Production

Hydrogen production pathways are very sensitive to hydrogen partial pressure. In the reactor, the hydrogen partial pressure rises as it accumulates in the headspace. The hydrogen is a byproduct of dark fermentation, which means that its accumulation would impede the reaction kinetics due to "Le Chatelier's" principle. The rise in partial pressure leads to metabolic shift resulting in formation of reduced end products such as acetone, butanol, lactate, propionate, and ethanol (Hawkes et al., 2007). The removal of hydrogen from the fermentation process was accomplished using a variety of methods such as negative displacement of water, membrane separation, flushing over head space with nitrogen, argon or methane. The hydrogen yield was improved by 68% on lowering partial pressure with occasional N_2 sparging (Mizuno et al., 2000). The hydrogen can be selectively absorbed from the overhead space of a reactor by a membrane composed of silane or polyvinyl trimethyl silane leading to reduction hydrogen partial pressure. This eventually led to improvement in hydrogen production (Teplyakov et al., 2002).

2.5.4 The Role of Hydraulic Retention Time (HRT) in Hydrogen Production

The HRT ("Hydraulic Retention Time") refers to the amount of time that the cells, as well as soluble nutrients, are resides within the reactor. The hydrogen production occurs at low HRTs than methanogenesis. The reactor volume and the feed flow rate determine hydraulic retention time (HRT=Volume of reactor/ Feed flow rate). The maximum rate of hydrogen production and substrate conversion can be achieved at specific HRT. A washout condition, in which all active cells leave the reactors due to low HRT, may also occur. Therefore, the optimization of HRT is linked to the organism's specific growth rate. On working with mixed consortia, which contain methanogens and acidogenic hydrogen producers, manipulating the HRT might lead to shift in microbial profile inside the reactor. The methanogens might get washed out of the reactor at lower HRTs, resulting in an increase in enrichment of acetogenic hydrogen producing (Lo et al., 2009). As a result, methanogens in mixed consortia might be totally suppressed by a pH acidity of 6–6.5 and a low HRT. Additionally, the HRT has a significant impact on the synthesis and breakdown of end metabolites. In case of *Clostridium* sp., the hydrogen production improved when HRT was reduced from 10 hours to 6 hours, followed by a decrement in propionate production (Zhang et al., 2006). On using anaerobic sludge, improvement in hydrogen production and the butyrate/acetate ratio was observed with HRT of one day with no trace of methane (Mariakakis et al., 2012). Studying HRT is useful when it comes to designing experiments and reactors for the treatment of industrial wastewater, where the adoption of low HRT by injecting transitory loading increased hydrogen production and also assisted in COD removal (Sentürk et al., 2013).

2.6 MATHEMATICAL MODELS EXPRESSING SUBSTRATE CONVERSION EFFICIENCY AND MICROBIAL GROWTH

Batch fermentation relies on the availability of a limiting substrate. The conventional Monod model is one of the commonly used unstructured mathematical models which describe microbial growth-associated substrate utilization in batch fermentation (Kargi and Shuler, 1992). The Monod equation Eq. 2.7 describing microbial growth is given below:

$$\mu = \frac{\mu_{max} S}{K_s + S} \tag{2.7}$$

Here, μ_{max} (h^{-1}) indicates maximum specific growth rate, "S" denotes the limiting substrate concentration (g L^{-1}), "K_s" denotes saturation constant (g L^{-1}) and μ(h^{-1}) represents specific growth rate. The Monod model may be linearized with Lineweaver–Burk formula. Regression analysis was used to determine the values of K_s and μ_{max}. Likewise, the kinetic constants mentioned above were calculated using complex sugars like starch as the limiting substrate for biohydrogen production. Lee et al. (2008) created a Monod model to illustrate the effect of a limited substrate (starch) on hydrogen generation (Eq. 2.8).

$$v_{H_2} = \frac{v_{max,H_2} C_{starch}}{K_s + C_{starch}} \tag{2.8}$$

Here v_{max} indicates the maximum rate of hydrogen production rate ($mLL^{-1}h^{-1}$), C_{starch} signifies initial starch concentration (g_{COD} L^{-1}) and K_s denotes half-saturation constant (g_{COD} L^{-1}). Using the conventional Monod model has a number of drawbacks. The impact of substrate inhibition during biohydrogen generation cannot be predicted by this model. The growth parameters cannot be successfully reproduced using the conventional Monod model when the substrate concentration was further increased to higher concentration. Furthermore, the Monod model did not account for the effects of other parameters like pH, cell density, substrate transport and metal presence. Various mathematical models with the above-mentioned characteristics were presented to mimic the growth and substrate utilization pattern. Numerous empirical models for studying substrate inhibition in pure or mixed culture for hydrogen production have been studied (Ntaikou et al., 2009).

Effect of substrate inhibition on hydrogen production has been studied using Andrew's model in comparison with the standard Monod model (Nath et al., 2008). The specific growth rate, as well as substrate concentration, were shown to have a nonlinear relationship (Eq. 2.9).

$$\mu = \frac{\mu_{max} S}{K_s + S + \frac{S^2}{K_i}} \tag{2.9}$$

Using Andrew's model (Eq. 2.10), the impact of varying the initial glucose content on biohydrogen production was studied in mixed culture (Wang et al., 2008). A significant regression coefficient (R^2) of 0.902 was found in the predicted kinetic constants.

$$r = \frac{67.1S}{47.7 + S + \frac{S^2}{13.5}} \tag{2.10}$$

Andrew's paradigm fails to capture the nature of substrate inhibition while accessing the impact of substrate inhibition. An enhanced Monod type model (Eq. 2.11) was developed, taking into account low-level substrate activation and high-level substrate inhibition (Han et al., 1998).

$$r = k(1 - \frac{S}{S_{max}})^n \times \frac{S}{S + K_s(1 - \frac{S}{S_{max}})^m} \tag{2.11}$$

Here, k indicates the hydrogen production rate constant (mL h^{-1}); S_{max} denotes the maximum substrate concentration (g L^{-1}), S signifies the substrate concentration (g L^{-1}) and r shows the rate of hydrogen production (mL h^{-1}). Production of hydrogen comes to a cease at S_{max}; K_s indicates saturation constant (g L^{-1}). The values of "m" and "n" may be calculated on the basis of the type of inhibition (mixed inhibition, uncompetitive, competitive, and noncompetitive). The substrate inhibition was discovered to be noncompetitive, with positive values for "m" and "n."

2.6.1 Mathematical Model Expressing Product Formation Kinetics

During the dark fermentative hydrogen production, two sorts of products are generated. The gaseous products are mostly hydrogen and CO_2, while the liquid products in the fermentation broth are metabolites generated during fermentation, like volatile fatty acids and solvents. Modified Gompertz formula (Eq. 2.12) is a widely used mathematical model for simulating product formation kinetics in biohydrogen generation.

$$H(t) = P\exp\left\{-\exp\left[\frac{R_m e}{P}(\lambda - t) + 1\right]\right\} \tag{2.12}$$

Here, P indicates the "gas production potential" (mL), Rm denotes the highest production rate (mLh^{-1}), λ signifies the lag time, t denotes incubation time(h), "e," represents the exp (1) 2.718 and H(t) indicates a hydrogen production cumulative volume (mL). The following equation may be used to simulate a nonlinear cumulative hydrogen production profile. The "H" varies as a function of time in batch fermentation. The maximum rate of hydrogen production (Rm) may be expressed as the linearly slope rising phase of hydrogen production (Wang et al., 2009).

For butyrate and acetate formation, the Gompertz model was used to get the kinetic constants (Mu et al., 2006). The following is the Gompertz formula (Eq. 2.13) for simulating acetate and butyrate formation:

$$P(t) = P_{max,i} \exp\left\{-\exp\left[\frac{R_{max,i}e}{P_{max,i}}(\lambda_i - t) + 1\right]\right\} \tag{2.13}$$

Here, P_i indicates the product "i" formed each liter of reactor working volume at time t; the potential highest product per liter of reactor working volume is expressed by $P_{max, i.}$ and "i" denotes butyrate or acetate.

In dark fermentation, the creation of hydrogen is deemed as a growth-related product. The Luedeking–Piret model may be used to understand this relationship mathematically (Obeid et al., 2009):

$$\frac{dP}{dt} = \frac{dx}{dt} + \beta x \tag{2.14}$$

Here β and α indicate growth and non-growth-associated product formation's coefficient respectively. Another approach, depending on a Luedeking–Piret modification, was suggested, in which biomass production and concurrent hydrogen generation were predicted based on growth-related nature (Eq. 2.15).

$$\frac{1}{x}\frac{dC_{H_2}}{dt} = \alpha\mu = \alpha\left[\frac{1}{x}\frac{dx}{dt}\right] \tag{2.15}$$

Here x denotes cell concentration (gVSS L^{-1}), C_{H2} indicates hydrogen concentration (mol), and μ signifies specific growth rate (h^{-1}). The slope reflects the "α" value when plotting C_{H2} rate vs biomass concentration. Using the aforementioned model, the "α" value for sucrose and xylose fermentation was 0.039 and 0.041mol g^{-1} Volatile Suspended Solids for *C. butyricum* (strains CGS5 and CGS2), and *C. pasteurianum* (strains CH7, CH5, CH4, and CH1), respectively ().

2.7 PROTOCOL FOR BIOHYDROGEN PRODUCTION

Study of biohydrogen production via a batch process in a customized bioreactor and measuring the following:

- Maximum specific growth rate (μmax)
- Saturation constants (K_S)
- specific growth rate(μ)

Materials required:

A) Microorganisms: Mesophilic facultative anaerobic hydrogen producers like *Enterobacter* sp.
B) Instruments and chemicals:
   - Customized bioreactors (500ml working volume)
   - Trap, CO_2 absorber, gas collectors, pipettes

- Gas chromatograms
- Spectrophotometer
- Centrifuge
- Dinitrosalicylic (DNS) solution
- Test tubes

C) Composition of the medium (Khanna et al., 2012)

Glucose	10g L^{-1}
Malt extract	10g L^{-1}
Yeast extract	4g L^{-1}
pH	6.5
Temperature	37°C

2.7.1 Experimental Setup for Dark Fermentation

The experimental setup is constructed as follows:

- An inoculum of 20% v/v seed culture is added to the bioreactor of 500ml working volume and allowed to incubate overnight (Figure 2.4).
- The temperature of the experimental system is maintained constant at 37°C using a circulating water bath.
- For the maintenance of an anaerobic environment in the reactor nitrogen gas (99.9% pure) is sparged.
- 40% w/v KOH solution is mixed to absorb the CO_2 produced during the reaction.
- The leftover gas (comprising of hydrogen mostly) is obtained in a gas collector by displacing 10percent w/v saline water at NTP.
- 5ml of the fermentation broth is collected at a specific time interval (hourly) aseptically and analysis of biomass and glucose concentration is determined of the same.

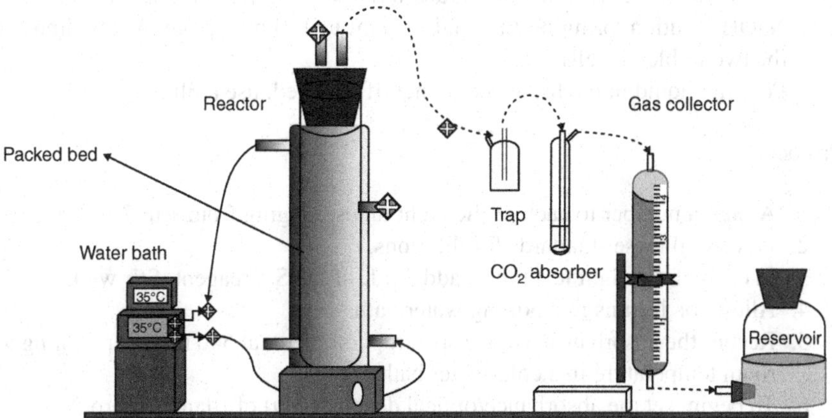

FIGURE 2.4 Dark fermentative hydrogen production setup.

2.7.2 Dry Cell Weight Determination

- The weight of the vacant Eppendorf is taken each time before taking the sample.
- All samples are centrifuged at 6000rpm for 15 minutes and the supernatant is separated to get the cell mass as a pellet.
- The pellets collected are washed with distilled water to remove traces of salts, minerals, and any media components present in the pellet after that the water is drained off.
- The collected pellet is again mixed with distilled water and centrifuged at 6000rpm for 10mins the mixture is stored overnight to dry at 60^0C.
- The Eppendorf tube with the pellet is weighed and the difference between the pellet containing Eppendorf and the empty Eppendorf is calculated to get the dry cell weight.

2.7.3 Glucose Estimation by Dinitro Salicylic Acid Method

Requirements

- Sodium potassium tartarate
- Sample
- Distilled water
- DNSA (Dinitro Salicylic Acid)
- 2 N sodium hydroxide (2N NaOH)

Procedure

- Make a 20mL solution of 2N NaOH.
- With the aid of a magnetic stirrer, dissolve weight 1g DNSA in 20mL NaOH.
- Dissolve 30 g sodium potassium tartarate in 50mL distilled water.
- Slowly add the sodium potassium tartarate solution into the DNSA and NaOH solution, bringing the total amount to 100 mL (Note: Allow time for the two to blend well).
- Pour the liquid into a brown container. If required, use a filter.

Protocol

1. Assign a number to each of the eight tubes, ranging from 1 to 7.
2. Prepare glucose standards for dilutions.
3. In each of the eight test tubes, add 3 mL of DNSA reagent. Stir well.
4. Allow for 15mins in a boiling water bath.
5. Record the absorbance using a spectrophotometer at 540 nm after cooling to room temperature in a cold-water bath.
6. To begin, set the absorbance/optical density (OD) of Blank to zero.
7. Take the OD of all the tubes and add them together (No. 1–7). After each dose of OD, wash the cuvettes.

Once biomass "x" and substrate "S" is estimated, the following equation will be used to plot ln (x/x_o) vs time.

$$ln\frac{x}{x_0} = \mu_{net}t \tag{2.16}$$

From the slope, the net specific growth rate(μ_{net}) may be calculated.

Similarly, the Eq. 2.17 mentioned below can help in estimating substrate Ks and μ_{max}.

$$\frac{1}{\mu} = \frac{Ks}{\mu_{max}}\frac{1}{S} + \frac{1}{\mu max} \tag{2.17}$$

For calculation of specific growth rate(μ) following equation should be followed:

$$\mu = \frac{1}{x_n}\frac{\{x_{n+1} - x_{n-1}\}}{2\Delta t} \tag{2.18}$$

CONCLUSIONS

Biohydrogen could emerge as a useful and sustainable alternative to fossil fuels as an energy. Dark fermentative pathway is the most sustainable as well as the eco-friendly method of producing biohydrogen, but its effectiveness must be greatly improved, and the cost of production must be dramatically decreased. When compared to conventional physicochemical techniques, the major expense of fermentative biohydrogen production is the feedstock. The cost of production could be reduced by using second- and third-generation biomass as feedstock. A plethora of microorganisms have been reported for biohydrogen production. Mesophilic species such as *Enterobacter, Citrobacter* etc., are a few well-known as hydrogen producers. Thermophilic microorganisms have shown greater yield and possess vivid enzymatic repertoire as compared to mesophiles. Furthermore, in-depth knowledge of biohydrogen production system biology, metabolic engineering, and genetic modification may ultimately lead to economically viable and competitive biohydrogen production. A greater emphasis must be given to the standardization of the operating parameters viz. pH, temperature, HRT, etc., to improve biohydrogen production.

REFERENCES

Antonopoulou, Georgia, Hariklia N. Gavala, Ioannis V. Skiadas, K. Angelopoulos, and Gerasimos Lyberatos. "Biofuels generation from sweet sorghum: fermentative hydrogen production and anaerobic digestion of the remaining biomass." *Bioresource Technology* 99, no. 1 (2008): 110–119.

Calli, Baris, Kim Schoenmaekers, Karolien Vanbroekhoven, and Ludo Diels. "Dark fermentative H_2 production from xylose and lactose—effects of on-line pH control." *International Journal of Hydrogen Energy* 33, no. 2 (2008): 522–530.

Daim, Tugrul, Diane Yates, Yicheng Peng, and Bertha Jimenez. "Technology assessment for clean energy technologies: The case of the Pacific Northwest." *Technology in Society* 31, no. 3 (2009): 232–243.

Fang, H., T. Zhang, and H. Liu. "Microbial diversity of a mesophilic hydrogen-producing sludge." *Applied Microbiology and Biotechnology* 58, no. 1 (2002): 112–118.

Ginkel, Steven Van, Shihwu Sung, and Jiunn-Jyi Lay. "Biohydrogen production as a function of pH and substrate concentration." *Environmental Science & Technology* 35, no. 24 (2001): 4726–4730.

Han, Keehyun, and Octave Levenspiel. "Extended Monod kinetics for substrate, product, and cell inhibition." *Biotechnology and Bioengineering* 32, no. 4 (1988): 430–447.

Hawkes, Freda R., Ines Hussy, Godfrey Kyazze, Richard Dinsdale, and Dennis L. Hawkes. "Continuous dark fermentative hydrogen production by mesophilic microflora: principles and progress." *International Journal of Hydrogen Energy* 32, no. 2 (2007): 172–184.

Jo, Ji Hye, Dae Sung Lee, Donghee Park, and Jong Moon Park. "Biological hydrogen production by immobilized cells of *Clostridium tyrobutyricum* JM1 isolated from a food waste treatment process." *Bioresource Technology* 99, no. 14 (2008): 6666–6672.

Jung, Gyoo Yeol, Jung Rae Kim, Ji-Young Park, and Sunghoon Park. "Hydrogen production by a new chemoheterotrophic bacterium *Citrobacter* sp. Y19." *International Journal of Hydrogen Energy* 27, no. 6 (2002): 601–610.

Kádár, Zsófia, Truus de Vrije, Giel E. van Noorden, Miriam AW Budde, Zsolt Szengyel, Kati Réczey, and Pieternel AM Claassen. "Yields from glucose, xylose, and paper sludge hydrolysate during hydrogen production by the extreme thermophile *Caldicellulosiruptor saccharolyticus*." In *Proceedings of the Twenty-Fifth Symposium on Biotechnology for Fuels and Chemicals Held May 4–7, 2003, in Breckenridge, CO*, pp. 497–508. Humana Press, Totowa, NJ, 2004.

Kalia, V. C., S. R. Jain, A. Kumar, and A. P. Joshi. "Frementation of biowaste to H_2 by *Bacillus licheniformis*." *World Journal of Microbiology and Biotechnology* 10, no. 2 (1994): 224–227.

Kamalaskar, Leena B., P. K. Dhakephalkar, K. K. Meher, and D. R. Ranade. "High biohydrogen yielding *Clostridium* sp. DMHC-10 isolated from sludge of distillery waste treatment plant." *International Journal of Hydrogen Energy* 35, no. 19 (2010): 10639–10644.

Kargi, Fikret, and Michael L. Shuler. *Bioprocess Engineering: Basic Concepts*. Prentice-Hall PTR (1992).

Khanal, Samir Kumar, Wen-Hsing Chen, Ling Li, and Shihwu Sung. "Biological hydrogen production: effects of pH and intermediate products." *International Journal of Hydrogen Energy* 29, no. 11 (2004): 1123–1131.

Khanna, Namita, Kanhaiya Kumar, Sona Todi, and Debabrata Das. "Characteristics of cured and wild strains of *Enterobacter cloacae* IIT-BT 08 for the improvement of biohydrogen production." *International Journal of Hydrogen Energy* 37, no. 16 (2012): 11666–11676.

Knappe, Joachim, and Gary Sawers. "A radical-chemical route to acetyl-CoA: the anaerobically induced pyruvate formate-lyase system of Escherichia coli." *FEMS Microbiology Reviews* 6, no. 4 (1990): 383–398.

Kotay, Shireen Meher, and Debabrata Das. "Microbial hydrogen production with Bacillus coagulans IIT-BT S1 isolated from anaerobic sewage sludge." *Bioresource Technology* 98, no. 6 (2007): 1183–1190.

Kumar, Narendra, Agnidipta Ghosh, and Debabrata Das. "Redirection of biochemical pathways for the enhancement of H2 production by *Enterobacter cloacae*." *Biotechnology Letters* 23, no. 7 (2001): 537–541.

Lee, Kuo-Shing, Ping-Jei Lin, and Jo-Shu Chang. "Temperature effects on biohydrogen production in a granular sludge bed induced by activated carbon carriers." *International Journal of Hydrogen Energy* 31, no. 4 (2006): 465–472.

Lee, Kuo-Shing, Yao-Feng Hsu, Yung-Chung Lo, Ping-Jei Lin, Chiu-Yue Lin, and Jo-Shu Chang. "Exploring optimal environmental factors for fermentative hydrogen production from starch using mixed anaerobic microflora." *International Journal of Hydrogen Energy* 33, no. 5 (2008): 1565–1572.

Lee, Yong-Eok, Mahendra K. Jain, Chanyong Lee, and J. Gregory Zeikus. "Taxonomic distinction of saccharolytic thermophilic anaerobes: description of *Thermoanaerobacterium xylanolyticum* gen. nov., sp. nov., and *Thermoanaerobacterium saccharolyticum* gen. nov., sp. nov.; reclassification of *Thermoanaerobium brockii, Clostridium thermosulfurogenes*, and *Clostridium thermohydrosulfuricum* E100–69 as *Thermoanaerobacter brockii* comb. nov., *Thermoanaerobacterium thermosulfurigenes* comb. nov., and *Thermoanaerobacter thermohydrosulfuricus* comb. nov., respectively; and transfer of …." *International Journal of Systematic and Evolutionary Microbiology* 43, no. 1 (1993): 41–51.

Levin, David B., Rumana Islam, Nazim Cicek, and Richard Sparling. "Hydrogen production by *Clostridium thermocellum* 27405 from cellulosic biomass substrates." *International Journal of Hydrogen Energy* 31, no. 11 (2006): 1496–1503.

Lo, Yung-Chung, Kuo-Shing Lee, Ping-Jei Lin, and Jo-Shu Chang. "Bioreactors configured with distributors and carriers enhance the performance of continuous dark hydrogen fermentation." *Bioresource Technology* 100, no. 19 (2009): 4381–4387.

Lu, Yuan, Liyan Wang, Kun Ma, Guo Li, Chong Zhang, Hongxin Zhao, Qiheng Lai, He-Ping Li, and Xin-Hui Xing. "Characteristics of hydrogen production of an Enterobacter aerogenes mutant generated by a new atmospheric and room temperature plasma (ARTP)." *Biochemical Engineering Journal* 55, no. 1 (2011): 17–22.

Mariakakis, Iosif, Carsten Meyer, and Heidrun Steinmetz. "Fermentative hydrogen production by molasses; effect of hydraulic retention time, organic loading rate and microbial dynamics." *Hydrogen Energy-Challenges and Perspectives* (2012): 121–148.

Mizuno, Osamu, Richard Dinsdale, Freda R. Hawkes, Dennis L. Hawkes, and Tatsuya Noike. "Enhancement of hydrogen production from glucose by nitrogen gas sparging." *Bioresource Technology* 73, no. 1 (2000): 59–65.

Mohr, Georg, Dorie Smith, Marlene Belfort, and Alan M. Lambowitz. "Rules for DNA target-site recognition by a lactococcal group II intron enable retargeting of the intron to specific DNA sequences." *Genes & Development* 14, no. 5 (2000): 559–573.

Morimoto, Kenji, Tetsuya Kimura, Kazuo Sakka, and Kunio Ohmiya. "Overexpression of a hydrogenase gene in *Clostridium paraputrificum* to enhance hydrogen gas production." *FEMS Microbiology Letters* 246, no. 2 (2005): 229–234.

Mu, Yang, Gang Wang, and Han-Qing Yu. "Kinetic modeling of batch hydrogen production process by mixed anaerobic cultures." *Bioresource Technology* 97, no. 11 (2006): 1302–1307.

Nandi, R., and S. Sengupta. "Microbial production of hydrogen: an overview." *Critical Reviews in Microbiology* 24, no. 1 (1998): 61–84.

Nath, Kaushik, Manoj Muthukumar, Anish Kumar, and Debabrata Das. "Kinetics of two-stage fermentation process for the production of hydrogen." *International Journal of Hydrogen Energy* 33, no. 4 (2008): 1195–1203.

Ngo, Tien Anh, Mi-Sun Kim, and Sang Jun Sim. "High-yield biohydrogen production from biodiesel manufacturing waste by *Thermotoga neapolitana*." *International Journal of Hydrogen Energy* 36, no. 10 (2011): 5836–5842.

Ntaikou, I., Hariklia N. Gavala, and G. Lyberatos. "Modeling of fermentative hydrogen production from the bacterium *Ruminococcus albus*: definition of metabolism and kinetics during growth on glucose." *International Journal of Hydrogen Energy* 34, no. 9 (2009): 3697–3709.

Obeid, Jamila, Jean-Pierre Magnin, Jean-Marie Flaus, Olivier Adrot, John C. Willison, and Roumen Zlatev. "Modelling of hydrogen production in batch cultures of the photosynthetic bacterium *Rhodobacter capsulatus*." *International Journal of Hydrogen Energy* 34, no. 1 (2009): 180–185.

Rainey, F. A., A. M. Donnison, P. H. Janssen, David Saul, A. Rodrigo, Peter L. Bergquist, Roy M. Daniel, E. Stackebrandt, and Hugh W. Morgan. "Description of *Caldicellulosiruptor saccharolyticus* gen. nov., sp. nov: an obligately anaerobic, extremely thermophilic, cellulolytic bacterium." *FEMS Microbiology Letters* 120, no. 3 (1994): 263–266.

Redwood, Mark D., Iryna P. Mikheenko, Frank Sargent, and Lynne E. Macaskie. "Dissecting the roles of *Escherichia coli* hydrogenases in biohydrogen production." *FEMS Microbiology Letters* 278, no. 1 (2008): 48–55.

Rosales-Colunga, Luis Manuel, and Antonio De León Rodríguez. "*Escherichia coli* and its application to biohydrogen production." *Reviews in Environmental Science and Bio/Technology* 14, no. 1 (2015): 123–135.

Roy, Shantonu, and Debabrata Das. "Ecobiotechnological Approaches: Enrichment Strategy for Improvement of H_2 Production." In *Microbial Factories*, pp. 29–45. Springer, New Delhi (2015).

Şentürk, Elif, Mahir İnce, and Guleda Onkal Engin. "The effect of transient loading on the performance of a mesophilic anaerobic contact reactor at constant feed strength." *Journal of biotechnology* 164, no. 2 (2013): 232–237.

Sinha, Pallavi, and Anjana Pandey. "An evaluative report and challenges for fermentative biohydrogen production." *International Journal of Hydrogen Energy* 36, no. 13 (2011): 7460–7478.

Stephenson, Marjory, and Leonard Hubert Stickland. "Hydrogenlyases: Bacterial enzymes liberating molecular hydrogen." *Biochemical Journal* 26, no. 3 (1932): 712.

Stickland, Leonard Hubert. "The bacterial decomposition of formic acid." *Biochemical Journal* 23, no. 6 (1929): 1187.

Taguchi, Fumiaki, Kiharu Yamada, Katsushige Hasegawa, Tatsuo Taki-Saito, and Kazuya Hara. "Continuous hydrogen production by *Clostridium* sp. strain no. 2 from cellulose hydrolysate in an aqueous two-phase system." *Journal of Fermentation and Bioengineering* 82, no. 1 (1996): 80–83.

Tanisho, S., and Y. Ishiwata. "Continuous hydrogen production from molasses by the bacterium *Enterobacter aerogenes*." *International Journal of Hydrogen Energy* 19, no. 10 (1994): 807–812.

Teplyakov, V. V., L. G. Gassanova, E. G. Sostina, E. V. Slepova, M. Modigell, and A. I. Netrusov. "Lab-scale bioreactor integrated with active membrane system for hydrogen production: experience and prospects." *International Journal of Hydrogen Energy* 27, no. 11–12 (2002): 1149–1155.

Tolvanen, K.E. and Karp, M.T., 2011. Molecular methods for characterizing mixed microbial communities in hydrogen-fermenting systems. *International Journal of Hydrogen Energy*, *36*(9), pp.5280–5288.

Van Groenestijn, J. W., J. H. O. Hazewinkel, M. Nienoord, and P. J. T. Bussmann. "Energy aspects of biological hydrogen production in high rate bioreactors operated in the thermophilic temperature range." *International Journal of Hydrogen Energy* 27, no. 11–12 (2002): 1141–1147.

Vatsala, T. M. "Hydrogen production from (cane-molasses) stillage by *Citrobacter freundii* and its use in improving methanogenesis." *International Journal of Hydrogen Energy* 17, no. 12 (1992): 923–927.

Wang, Jianlong, and Wei Wan. "Optimization of fermentative hydrogen production process using genetic algorithm based on neural network and response surface methodology." *International Journal of Hydrogen Energy* 34, no. 1 (2009): 255–261.

Wang, Jianlong, and Wei Wan. "The effect of substrate concentration on biohydrogen production by using kinetic models." *Science in China Series B: Chemistry* 51, no. 11 (2008): 1110–1117.

Wang, Ruofan, Wenming Zong, Changli Qian, Yongjun Wei, Ruisong Yu, and Zhihua Zhou. "Isolation of *Clostridium perfringens* strain W11 and optimization of its biohydrogen production by genetic modification." *International Journal of Hydrogen Energy* 36, no. 19 (2011): 12159–12167.

Weizmann, Chaim, and Bruno Rosenfeld. "The activation of the butanol-acetone fermentation of carbohydrates by *Clostridium acetobutylicum* (Weizmann)." *Biochemical Journal* 31, no. 4 (1937): 619.

Wiegel, Jürgen, and Lars G. Ljungdahl. "*Thermoanaerobacter ethanolicus* gen. nov., spec. nov., a new, extreme thermophilic, anaerobic bacterium." *Archives of Microbiology* 128, no. 4 (1981): 343–348.

Zeidan, Ahmad A., and Ed WJ Van Niel. "Developing a thermophilic hydrogen-producing co-culture for efficient utilization of mixed sugars." *International Journal of Hydrogen Energy* 34, no. 10 (2009): 4524–4528.

Zhang, Zhen-Peng, Kuan-Yeow Show, Joo-Hwa Tay, David Tee Liang, Duu-Jong Lee, and Wen-Ju Jiang. "Effect of hydraulic retention time on biohydrogen production and anaerobic microbial community." *Process Biochemistry* 41, no. 10 (2006): 2118–2123.

3 Biomethane Production Process

3.1 INTRODUCTION

Volta was the first to describe methane generation from water-saturated decomposing organic plant materials. Nearly a century later, the relation of methane production with microbes was established. In 1882, Tappeiner provided more adequate evidence of methane's microbiological origin. Three similar anaerobic cultures were given with the organic part of plants as substrate and intestinal content of ruminants as the seed culture. The rumen of herbivorous animals' harbors cellulose degrading organisms but the role of such organisms towards methane production was not clear to the animal physiologist. With the understanding of syntrophism among various anaerobic microorganisms, a clear vision of the complex process of methane production was obtained.

3.2 MICROBIOLOGY AND BIOCHEMISTRY OF METHANOGENESIS

3.2.1 Taxonomic Diversity of Methanogens

The defining characteristic of methanogens is their ability to produce methane and other hydrocarbons which set them apart from other microbes. Methanogens belong to *Archaeobacteria* which are different from Eubacteria in many aspects like the existence of membrane lipids with isoprenoids connected to glycerol by ether linkage, the lack of peptidoglycan comprising muramic acid, and distinct RNA sequences of ribosomal (Balch et al., 1979; Raskin et al., 1994). Methanogens may be categorized into three types on the basis of their metabolic processes for producing methane: aceticlastic, methylotrophic, and CO_2-reducing. In the CO_2-reducing methanogenic microorganisms, a two-electron reduction is required to convert bicarbonate or CO_2 to methane (Rouviere and Wolfe, 1988). The H_2 is the only source of electrons for many methanogens. The source of H_2 in natural habitats might be geological eruptions or H_2 created by other hydrogen-producing bacteria. Under anaerobic conditions, the H_2 is rapidly consumed by methanogens

DOI: 10.1201/9781003224587-3

so it never gets accumulated in the system. Therefore, H_2 serves as a key role of extracellular intermediate.

Formate is used as an electron donor by several hydrogenotrophic methanogens for CO_2 reduction to CH_4. Similar to H_2, formate is also an important intermediate in methane production although its level in the methanogenic environment remains low (Boone et al., 1989). Very few methanogens also have the capability of oxidizing primary as well as secondary alcohols in order to reduce CO_2 to CH_4 (Bleicher et al., 1989). On the other hand, methanol, dimethyl sulfide, and trimethylamine are examples of methylotrophic methanogens that can employ methyl groups as a substrate. Methane is decreased once the methyl group is moved to a methyl carrier (coenzyme M). Initially, the methanogenic bacterial species were categorized along with non-methanogens based on morphological properties. The 8th edition of *Bergey's Manual* recognized the physiological unity of methanogens and classified them as a single category (Bryant and Boone, 1987). Further methanogens unity has been indicated with the advent of the ribotyping technique which comprises sequencing and cataloging of 16S rRNA. Amazingly, methanogens showed phylogenetic relation with some severe halophiles as well as thermophilic sulfur-dependent organisms from kingdom *Archaeobacteria*. The methanogens as well as other *Archaeobacteria* were categorized under a novel taxonomic higher level kingdom termed as kingdom "*Archaea*" (Woese et al.,1990). Under the kingdom "Archaea," the methanogens have been involved in the "Euryarchaeota." The "*Euryarchaeota*," also *involves Thermoplasma*, extreme halophiles, and some nonmethanogenic thermophilic extremophiles.

3.2.2 Taxonomical Classification of Methanogens

Within the kingdom *Archaeobacteria*, methanogens may be categorized into five orders. The order *Methanomicrobiales*, *Methanococcales*, and *Methanobacteriales* were the three distinct methanogens described in "*Bergey's Manual of Systematic Bacteriology*" (Mah and Kuhn, 1984) (Figure 3.1). The aceticlastic methanogens and methylotrophic have been further separated from the order of *Methanomicrobiales* and reordered under *Methanosarcinales* (Bélaich et al., 2012). Moreover, a new order of *Methanopyrales* was discovered which was phylogenetically distinct from all the known methanogens (Burggraf et al., 1991).

3.2.2.1 Methane-producing *Methanobacteriales* sp.

Methanobacteriaceae and *Methanothermaceae* are two families of the *Methanobacteriales* order. *Methanobacteriales* are mainly rod-shaped methanogens that utilize CO_2 as an energy source. The only exception of the above group is *Methanosphaera sp* which is shaped like a cocci and uses H_2 as an energy source to convert methanol to methane. The *Methanobacteriales* strains are generally gram-positive because they have pseudomurein cell walls.

The *Methanobacteriaceae* is a large and varied family of bacteria. It comprises numerous genera like *Methanothermobacter* gen. nov., *Methanobacterium*,

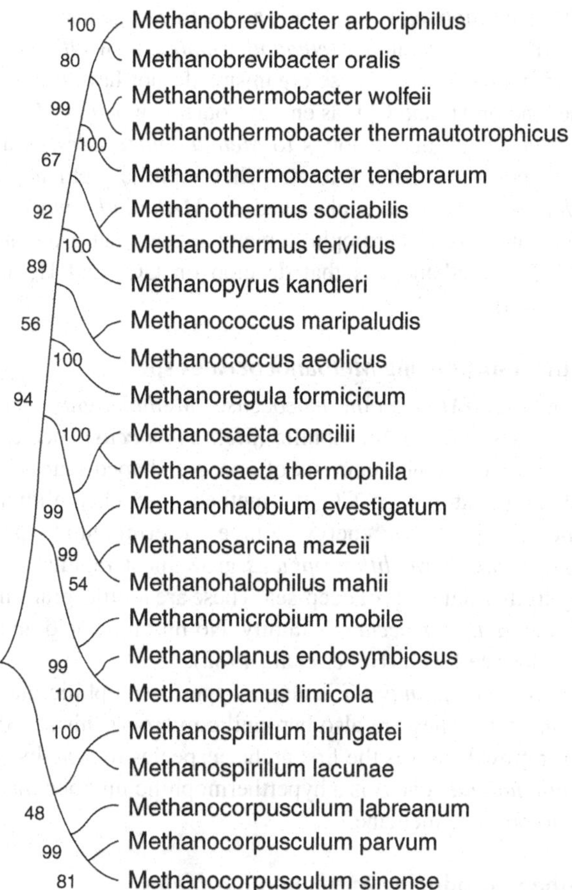

FIGURE 3.1 Phylogenetic tree of methanogens.

Methanosphaera, and *Methanobrevibacter. Methanobacterium jormicicum* is the earliest known *Methanobacterium* species. *Methanobacterium bryantii* differs from *M. jormicicum* in terms of physiological and morphological characteristics, such as the latter's inability to catabolize formate. *Methanobrevibacter sp.* is usually discovered in the gastrointestinal tract or excrement of mammals. They have complicated organic needs and are shaped like short rods or cocco-bacilli. To convert CO_2 to CH_4, they need H_2 or formate as an energy source.

Another cocci-shaped member of *Methanobacteriaceae* is *Methanosphaera*. They are non-motile, gram-positive, exist individually or in tiny groups, and grow by converting methanol (CH_3OH) to methane with the help of H_2. Serine may be found in the pseudomurien cell wall. Two members of this genus are *Methanosphaera stadmaniae* and *M. cuniculi* (Biavati et al., 1988).

The *Methanothermobacter* sp. is a distinct genus of thermophilic *Methanobacteriales*. They contain *Methanothermobacter wollfi and Methanothermobacter thermoautotrophicus*. These organisms do not have the ability to utilize formate and depend on H_2 and CO_2 as energy sources, whereas *Methanobacterium thermoformicicum*, which also belongs to *Methanobacteriales*, is a thermophilic methanogen capable of utilizing formate. Thermophilic methanogens belonging to order *Methanobacteriales* are classified as *Methanothermaceae*. The genus *Methanothermus* are severe thermophilic methanogens that grow at temperatures between 83–85°3C. Methanogens that develop on CO_2 and H_2 are rod-shaped (Lauerer et al., 1986).

3.2.2.2 Methane-producing *Methanococcales* sp.

Three thermophilic (*Methanothermococcus*, *Methanoignis*, and *Methanocaldococcus*) species and one mesophilic (*Methanococcus*) species make up the order Methanococcales, which consists of coccoid-shaped marine methanogens. Halophilic bacteria that reduce CO_2 to methane are chemolithotrophic. They utilize formate or H_2 as an energy source. Thermophilic species such as *Methanothermococcus thermolithotrophicus* grow most rapidly at 65°C and are identified in hydrothermal vents of deep-sea. These are motile, gram-negative cocci, that belong to the *Methanococcaceae* family. No muramic acid and glycoprotein are observed in their cell wall (Huber et al., 1982).

Methanocaldococcus jannaschii is also a hyperthermophilic marine cocci that grows rapidly at 85°C. They are also barophilic microorganism and requires 200 atm pressure for growth. It was the first archaea species to have its whole genome sequenced. *Methanoignis igneus* is a hyperthermophilic microbe that uses H_2 as an energy source to produce methane.

3.2.2.3 Methane-producing *Methanomicrobiales*

It consists of three families, namely *Methanocorpusculaceae*, *Methanosarcinaceae*, and *Methanomicrobiaceae. Methanomicrobiales* species require acetate as a carbon source for their cells. Several other species have more complicated nutritional needs. This order also contains thermophilic or mesophilic microbes that are also slightly halophilic. These organisms' cell walls have S-layer (protein layer). An external sheath is also existing in some species such as *Methanospirillum hungateii*. The presence of S-layer makes microorganisms of these groups osmotically sensitive. They are vulnerable to hypotonic shock and dilute detergents like 2% v/v SDS. A Pleomorphic, irregular coccoid is the most common shape. Many have a rod or plate-shaped cells. *Methanospirillum hungateii* has a helical spiral shape which is covered by the sheath. The "*Methanocorpusculaceae*" are hydrogenotrophic methanogens *(Methanocorpusculum)* with a cocci form. They convert CO_2 to methane by oxidizing H_2, formate, or alcohol. The *Methanomicrobium mobile* and *Methanolacinia paynteri* is the single representative of their group.

Methanoplanus contains two types: *Methanoplanus endosymbiosus* and *Methanoplanus limicola*. These microorganisms are weakly motile and are plate shaped. The envelope of the cells has a hexagonal configuration. These were found in swampy marshes. They are gram-negative acetoclastic microorganisms and utilize formate and H_2 to create methane. The genus *Methanoculleus* comprises four mesophilic species viz. *M. olentangyi, M. thermophilicus, M. marisnigri*, and *Methanoculleus bourgensis* which utilize formate/H_2 as methanogenic substrates (Asakawa, 2003).

3.2.2.4 Methane-producing *Methanosarcinales*

It may be categorized as *Methanohalobium, Methanohalophilus, Methanococcoides, Methanolobus, Methanosaeta (Methanothrix)*, and *Methanosarcina. The nutritional requirements of the members of Methanosarcinaceae are* methyl-group comprising compounds like methyl sulfides, methylamines, or methanol thus they are called methylotrophic microorganisms. The microorganisms belonging to these groups can use trimethylamine and dismutates to produce CH_4, ammonia, and CO_2. Similarly, methanol could also catabolize to CO_2 and methane. Some *Methanosarcinaceae* species are acetoclastic or hydrogenotrophic thus may convert CO_2 to CH_4 or use acetate to produce CO_2 and methane.

Another distinctive aspect of *Methanosarcinaceae* is that they are methylotrophic and none of them may utilize formate as a "catabolic substrate." The aceticlastic genus *Methanosaeta (Methanothrix)* is found only in the family Methanosaetaceae. They are gram-negative, non-motile rods with flat ends, ranging in length from 2.5 µm to 6.0 µm. *Methanosaeta* utilizes acetate as an energy source through aceticlastic reactions. No other substrate supports its growth other than acetate. A sheath-like structure confers a rod-shaped structure to these organisms and the organisms grow within this sheath. They often create a lengthy chain structure.

3.2.2.5 Methane-producing *Methanopyrales* ord. nov.

Methanopyraceae has been categorized as a distinct methanogen group. The genus *Methanopyrus* comprises a single species: M. *kandleri*. It generally has a rod-shape, gram-positive, and develops at very high temperatures. Their cell wall is made up of a special sort of pseudomurein that includes ornithine in addition to lysine but doesnot include "N-acetylglucosamine" (Kurr et al., 1991). They are hydrogenotrophic bacteria that convert CO_2 into methane.

3.3 MICROBIAL INTERACTIONS

3.3.1 General Considerations on Competition/or Methanogenic Substrates

After acetogenic hydrogen production, the spent medium is rich in volatile fatty acids like butyrate, acetate, ethanol, etc. On adjusting the pH of the spent media to the alkaline range (pH 7.2 to pH 8). There are three major bacteria groups that compete with methanogens for the substrate. The methanogenic seed cultures generally have metal-reducing bacteria, sulfate-reducing bacteria as co-contaminants.

Gram-negative proteobacteria are the bacteria that reduce sulfate. In comparison to methanogens, it may utilize a significantly wider range of electron donors. Sulfate-reducing bacteria may use aromatic chemicals, amino acids, alcohols, and organic acids, as electron donors. Hydrogenotrophic microorganisms are gram-positive eubacterial and could also utilize a variety of substrates, like purines, sugars, as well as methoxyl groups of methoxylated aromatic compounds.

A hierarchy for electron donor competition is seen in a setting where the organic substrate (electron donor) is limited. If there are potential electron acceptors in the system, metal-reducing bacteria like Fe^{3+} reducers may outcompete other species. A succession of sulfate-reducing microorganisms, acetogens, and methanogens would follow. When molecular hydrogen was used as the electron donor, a 12.5% higher $\Delta G^{O'}$ was observed for sulfate reduction as compared to methanogenesis. Thus, in settings with high sulfate concentrations, methanogenesis is entirely inhibited.

3.3.1.1 Competition for Hydrogen

After completion of 1st stage where hydrogen production takes place, a significant quantity of H_2 remains in the overhead area and also in the dissolved form. The molecular H_2 is consumed under methanogenic conditions (2nd stage). For estimating the H_2 utilization rate via hydrogenotrophic methanogens, the H_2 consumption rate must be slow enough that the process is not slowed with the H_2 transition from the gas state to a liquid state. The competition for H_2 may be investigated by examining the apparent Km values for H_2 use under anaerobic conditions. The methanogens and methanogenic habitats have apparent Km values of *4–8* μM H_2 (550 Pa to1100 Pa). The values are even lower in the case of sulfate-reducing bacteria, about *2* μM (Table 3.1).

The higher Km values are most likely due to the uptake hydrogenases' inherent limitations in utilizing H_2 at lower partial pressures. The effect of partial pressure of H_2 could be observed in a bioreactor with a greater loading rate.

TABLE 3.1
H_2 Intake by Pure Cultures and Methanogenic Settings Based on Apparent K_m Values

	Apparent K_m		
Organism/ Habitat	**μM**	**Pa**	**References**
Desulfovibrio formicicum	2	250	Kristjansson et al., 1982
Sewage sludge	4–9	740	Robinson et al., 1982
Rumen fluids	4–9	860	Robinson et al., 1982
Methanobacterium thermoautotrophicum	8	1,700	Kristjansson et al., 1982
Methanosarcina barkeri	13	1,700	Kristjansson et al., 1982
Methanospirillum hungatei	5	670	Robinson et al., 1982
Desulfovibrio vulgaris	6	800	Robinson and Tiedje 1984

Another approach that could be used to recognize anaerobes competition for H_2 is by correlating free energy available with H_2 threshold partial pressure. The H_2 thresholds values were measured in a variety of hydrogenotrophic anaerobes (Cord-Ruwisch et al., 1988).

The reaction's free energy and hydrogen threshold were observed to have an inverse relationship (Table 3.2). The order of the threshold value is acetogen>methanogens>sulfate reducers. This means that sulfate-reducing bacteria may lower the partial pressure of H_2 to a point where even methanogens are unable to utilize it. These threshold values are reaction-specific, not organism-specific. The thermodynamic influence of H_2 partial pressure on H_2 uptake rate may be described with free energy calculation using the Nernst formula. For a chemical process that takes place at 25°C:

$$\mathrm{aA + bB \rightarrow cC + dD} \tag{3.1}$$

The values of ΔG' may be approximated (pH = 7) in kJ as stated in Eq. 3.2.

$$\Delta G' = \Delta G^{O'} + RT \ln \frac{(C)^c (D)^d}{(A)^a (B)^b} = \Delta G^{O'} + 5.7 \log \frac{(C)^c (D)^d}{(A)^a (B)^b} \tag{3.2}$$

Here "*R*" signifies the ideal gas constant, "*T*" indicates the absolute temperature in kelvin and "A" represents the reactant A's molar concentration.

The free energy may be represented by the Nernst equation (Eq. 3.3) for the methanogenesis process that involves H_2-CO_2 reduction. Assuming HCO_3 concentration 10mM as well as methane partial pressure of 0.5atm, dependency on free energy (ΔG') may be computed as follow:

$$\Delta G' = -131 + 5.7 \log \frac{(CH_4)}{(HCO_3^{1-})} - 5.7 \log (H_2)^4 = -123 - 22.8 \log (H_2) \tag{3.3}$$

3.3.1.2 Competition for Acetate

Very little information is available in terms of the physiological characteristics of hydrogenotrophic methanogen species. Properties that would favor one hydrogenotrophic species over another are still debatable. But in the case of the role of acetate in methanogenesis, many reports are available. A higher concentration of acetate favors rapid growth of *Methanosarcina* sp. *On the contrary*, members of the *Methanothrix sp. favor low acetate contents*. The microbial profile showed *Methanothrix* dominance when the concentration of acetate in a thermophilic anaerobic digester reduced below 1mM (Zinder et al., 1984). To describe the competition for acetate, both Michaelis-Menten, as well as threshold models, have been used (as depicted in the instance of hydrogenotrophic microorganisms). For acetoclastic methanogens, the minimal threshold for acetate usage studied for *Methanosarcina*, is generally 0.5mM and higher, but for "*Methanothrix*" are in the micromolar range. Methanogenesis is more likely to occur at slightly alkaline pH

TABLE 3.2
Threshold for Different Hydrogenotrophic Microorganisms

Microorganisms	Electron Accepting reaction	$\Delta G^{o\prime}$ (kJ/mol H_2)	H_2 threshold (Pa)
Methanospirillum hungatei	$CO_2 \rightarrow CH_4$	–33.9	3
Methanobrevibacter smithii	$CO_2 \rightarrow CH_4$	–33.9	10
Acetobacterium woodii	$CO_2 \rightarrow$acetate	–26.1	52
Desulfovibrio desulfuricans	$CO_2 \rightarrow H_2S$	–38.9	0.9

Source: Cord-Ruwisch et al., 1988.

ranges (7.2 to 8.0). Acetate dissociates poorly at those pH ranges. Such threshold values are caused by undissociated acetic acid at such pH levels.

3.3.1.3 Obligate Interspecies Hydrogen/ Formate Transfer

Alcohols, butyrate, and propionate were formerly considered to allow methanogens to grow for longer periods than methanol.

An anaerobic methanogenesis highlight is a symbiotic association between methanogens. It contains symbiotic bacteria that oxidize ethanol to acetate, and *Methanobacterium* will decrease CO_2 to CH_4 using an electron from hydrogen. By the occurrence of syntrophisim, it was also established that a hydrogen-producing microbe and a hydrogen oxidizing organism may co-exist via breaking a single substrate. Therefore, physical proximity in hydrogen producers and users might aid hydrogen transmission (Conrad et al., 1985). The *Syntrophomonas wolfei* when co-cultured with "*Methanospirillum hungatei*," grows faster in the existence of formate. But when *Syntrophomonas wolfeiis* was co-cultured with *Methanobacterium bryantii* which cannot use formate, it grew slowly (McInerney et al., 1981). This emphasizes the fact that physiological variations between two genera could also have a vital role in the two-stage bio methanation system. An interesting observation was made in thermophilic methanogenesis. Under thermophilic conditions, acetate and propionate oxidizing bacteria might combine with *Methanobacterium thermoautotrophicum* (which may use formate). The H_2 partial pressure at high temperature should be higher as well as high temperature also aids diffusion. Therefore, formate concentration may not be as important under thermophilic conditions.

3.3.1.4 Interspecies Acetate Transfer

Under anaerobic conditions, the accumulation of acetate is countered by acetoclastic methanogens thereby playing a vital role in keeping pH homeostasis inside the reactor. Acetate is a primary result of syntrophic processes. The thermodynamics of methane generation might be affected by acetate accumulation. Syntrophic degradation of butyrate yields 2 moles of hydrogen and acetate respectively. Thus, on reaction thermodynamics, a ten-fold increase in acetate level has the same impact as hydrogen. *Methanosarcina barkeri* bioaugmentation to a "*Syntrophomonas wolfei-Methanospirillum hungatei*" coculture facilitates butyrate degradation (McInerney and Beaty, 1988). In general, acetate concentration is usually higher as compared to the dissolved concentration of H_2. Therefore, acetate turnover is unlikely to be affected as much as H_2 turnover. The acetone-degrading methanogenic enrich culture might be a case of obligatory interspecies acetate transfer (Schink and Platen, 1987). The filamentous *Methanothrix sp.* dominated the mixed culture when acetone was supplied as the only carbon and energy source. Acetone was converted to acetoacetate by carboxylation. After splitting into 2 moles of acetate, the acetoacetate is transformed into methane. Acetone degradation was inhibited when acetate was added externally to the system. On addition of BES (bromoethane sulfonate) which is an inhibitor of methanogenesis also inhibited acetone degradation. The inference drawn from the above observation was that the ΔG for converting of "acetone to acetate" was -34.2kJ per reaction, the effectiveness of acetone-degrading microorganisms was based on acetate degradation.

3.4 BIOCHEMISTRY OF BIOMETHANE PRODUCTION

The spent media-generated after dark fermentative H_2 production are rich in SCFA ("Short Chain Fatty Acids") like acetate, butyrate, propionate, and lactate. These SCFAs might be a potential substrate for methanogenic microorganisms. Furthermore, the dissolved CO_2 and H_2 in the spent medium could also serve as a substrate for a distinct group of methanogens (hydrogenotrophic methanogens). Interactions among different groups of acetoclastic and hydrogenotrophic methanogens are greatly influenced by the action of many metabolic enzymes.

3.4.1 Methanogenesis from CO_2 and H_2: Bioenergetic of Hydrogenotrophic Methanogens

Gaseous compounds like CO_2 and H_2 are formed during acidogenic dark fermentation. A substantial amount of dissolved CO_2 and H_2 remains inside the spent medium. These dissolved gases might be used as a source of energy in the metabolic process of some kind of methanogen (Figure 3.2). The CO_2 and H_2 are the only sources of energy for methanogenic microorganisms (Balch et al., 1979), but there exist a few exceptions like *Methanothrix* spp., that can perform only acetate metabolism (Huser et al., 1982); *Methanolobus tindarius*, which could only use methanol as well as methylamine (König and Stetter, 1982); and *Methanosphaera stadtmaniae*, which is liable for reduction of methanol with H_2 (Miller and Wolin, 1985).

$$CO_2 + 4H_2 \rightarrow CH_4 + 2H_2O \; \left(\Delta G^{o} = -131 \text{ kJ mol}^{-1}\right) \tag{3.4}$$

The partial pressure of hydrogen in methanogenic ecosystems is typically between 1Pa and 10Pa, with the related free energy change (ΔG') between CO_2 and H_2

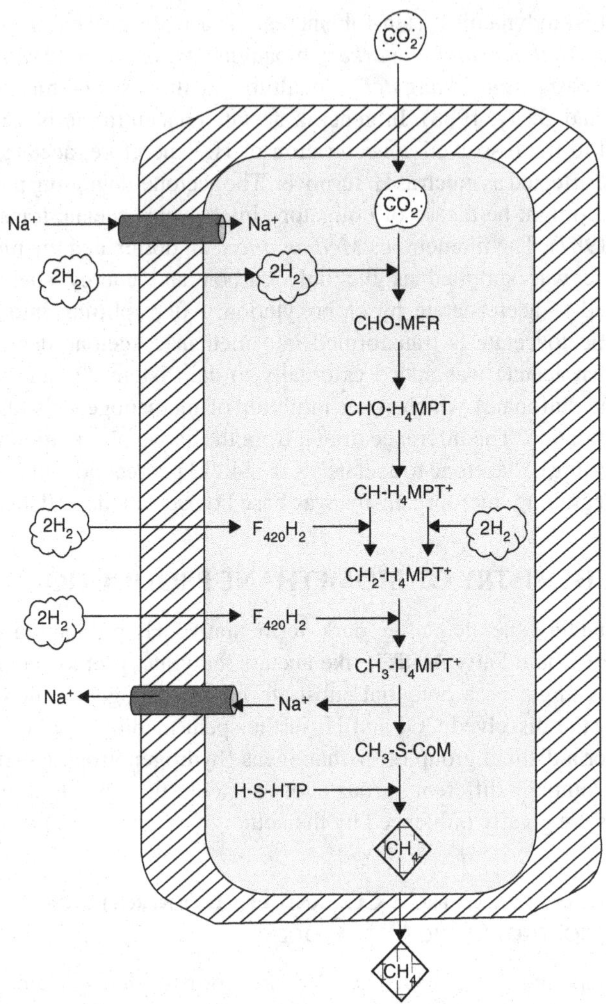

FIGURE 3.2 Methane production by hydrogenotrophic methanogens. CHO-MFR, N-formylmethanofuran; CHO-H_4MPT, N^5-formyltetrahydromethanopterin; CH-H_4MPT^+, N^5 -N^{10}-methenyltetrahydromethanopterin; CH_2-H4MPT, N^{10}-methylene tetrahydromethanopterin; CH_3-H_4MPT^+, N^5-methyl tetrahydromethanopterin; CH_3-S-CoM, methyl-coenzyme M; H-S-HTP, N-7-mercaptoheptanoyl-O-phospho-L-threonine; $F_{420}H_2$ reduced coenzyme F_{420}.

ranging from -20kJ mol^{-1} to -40kJ mol^{-1}. Though, in vivo ATP production from inorganic phosphate and ADP, necessitates at least 50kJ mol^{-1} free energy (Thauer and Morris, 1984). Under normal physiological settings, lower than 1mol ATP mol^{-1} CH_4 may be created throughout growth. Several studies have proposed the presence of a chemiosmotic process that couples exergonic CH_4 production with endergonic ADP phosphorylation (Blaut et al., 1990).

3.4.2 Acetoclastic Methanogenesis

Earlier research in the field of acetate conversion to methane has various theories for a description of the overall mechanism of the process. In 1936, the evidence of reduction of CO_2 to CH_4 along with the generation of $4H_2$ molecules by the acetate oxidation to two CO_2 molecules was found (Barker, 1936). Later, with the use of ^{14}C-labeled acetate, it was established that the carboxyl carbon, as well as methyl group, were major contributors to methane production. The CO_2 reduction hypothesis was ruled out by this observation. It was also found out in further studies that the methyl group's hydrogen (deuterium) atoms are directly converted to CH_4. Additional studies into methyl group conversion in different substrates led to the conclusion that reductive demethylation of a general precursor X-CH3 is the last phase in methanogenesis from all substrates (Pine and Vishniac, 1957). Reduction of acetate to CH_4 by the supply of exogenous hydrogen was another theory that developed over time. Nevertheless, this process was ruled out by observing that acetate is the only source of growth as well as methanogenesis in pure cultures. In most of the acetate-using bacterial anaerobes, cleavage of acetyl-CoA is followed by a drop of an external electron acceptor and oxidation of the carbonyl and methyl groups to CO_2.

The acetotrophic methane-producing "Archaea" group also cleaves acetate. Though, in this case, methane is formed by decreasing the methyl group using electrons obtained from the carbonyl group's oxidation to CO_2. Thus, it can be concluded that the acetate conversion to CO_2 and CH_4 is a fermentative process. The phylogenetic difference between the two areas creates doubts for a valid comparison of biochemical mechanisms involved in the cleavage and activation of acetate. Figure 3.3 shows the process as it is now understood in methane production from acetate. It is generally observed in *Methanosarcina* sp. and *Methanothrix* sp. As described, the first step in the pathway is the acetate activation into acetyl-CoA which is proceeded by cleavage of the carbon-sulfur bonds (decarbonylation) and carbon to carbon. It is catalyzed with the nickel/iron-sulfur component of the "CO dehydrogenase" (CODH) enzyme complex. Enzyme components containing nickel/iron-sulfur are responsible for both CO_2 reduction and carbonyl group oxidation to CO_2 in this process. At least two methyltransferases catalyze the methyl group transfer from the iron-sulfur/corrinoid component inside the complex to coenzyme M(HS-CoM). The heterodisulfide "CoM-S-S-HTP" is formed when "CH_3-S-CoM" is demethylated to CH_4 using electrons produced by sulfur atoms in HS-HTP in CH_3-S-CoM. The electrons obtained from reduced ferredoxin are used to reduce the "heterodisulfide" to the corresponding cofactors sulfhydryl forms; however, the process requires further investigation as the electron transfer mechanism from ferredoxin into the heterodisulfide is still unidentified.

3.4.2.1 Acetate Activation to Acetyl-CoA

The need for "active acetate" prior to cleavage. Acetyl phosphate and acetyl- CoA were found to be the potentially activated forms of acetate for methane formation.

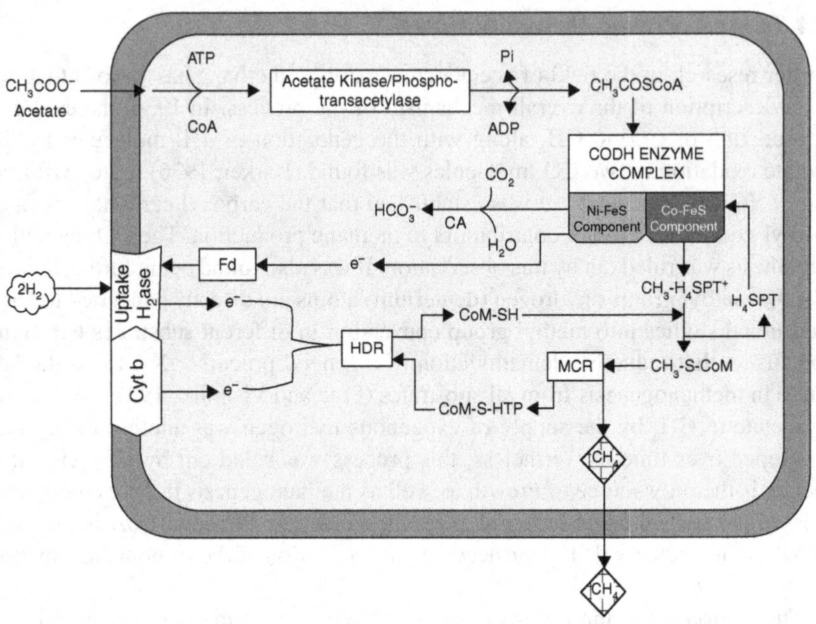

FIGURE 3.3 Acetogenic methanogenesis H_4SPT, tetrahydrosarcinapterin; CA, carbonic anhydrase; MCR, active methylreductase; HDR, heterodisulfide (CoM-S-S-HTP) reductase; Fd, ferredoxin; CODH, carbon monoxide dehydrogenase; cyt b, cytochrome b; H_2ase, hydrogenase.

Only when grown on acetate can certain microorganisms, like *M. thermophila*, produce acetate kinase as well as phosphotransacetylase (Lundie and Ferry, 1989), a conclusion that is compatible with acetate activation through reactions 3.5 and 3.6.

$$CH_3COO^- + ATP \rightarrow CH_3CO_2PO_3^{2-} + ADP \ (\Delta G^{0'} = 13 kJ/mol) \tag{3.5}$$

$$CH_3CO_2PO_3^{2-} + CoA \rightarrow CH_3COSCoA + Pi \ (\Delta G^{0'} = -9 kJ/mol) \tag{3.6}$$

$$CH_3COSCoA + H_4SPT \rightarrow CoA + CO + CH_3\text{-}H_4SPT \ (\Delta G^{0'} = 62 kJ/mol) \tag{3.7}$$

$$CH_3\text{-}H_4SPT + HS\text{-}CoM \rightarrow CH_3\text{-}S\text{-}CoM + H_4SPT \ (\Delta G^{0'} = -29 kJ/mol) \tag{3.8}$$

$$CH_3\text{-}S\text{-}CoM + HS\text{-}HTP \rightarrow CH_4 + CoM\text{-}S\text{-}S\text{-}HTP (\Delta G^{0'} = -43 kJ/mol) \tag{3.9}$$

Interestingly, the half-maximal rates (K_m) values of acetate kinase are lower for *Methanothrix* sp. (K_m = 0.86 mM acetate) as compared to *Methanosarcina* sp. (K_m = 22 mM acetate) (Jetten et al., 1990). This is compatible with Methanothrix's capacity to outcompete *Methanosarcina* for low acetate concentrations in the environment (Jetten et al., 1990). For both *Methanosarcina* and *Methanothrix* sp., expenditure of one ATP equivalent is made for each mole of acetate metabolized. *M. thermophila* falls into the category of strict anaerobes but has an air-stable enzyme acetate kinase which has a subunit of 53,000 Da. Also, phosphotransacetylase, another air-stable monomeric protein present in *M. thermophila* (Mr = 42,000 Da) is isolated from the soluble fraction (Lundie and Ferry, 1989). The occurrence of ammonium or potassium ions in concentrations of around 50mM increases the enzyme activity by seven-fold. Moreover, ions such as phosphate, arsenate, and sulfate cause inhibition of this enzyme. Similar to acetate kinase obtained from *M. thermophila, M. soehngenii* also has an air-stable enzyme acetyl-CoA synthetase. The special characteristic of these enzymes is in the presence of an ATP binding site.

3.4.2.2 Cleavage of the Carbon-Sulfur and Carbon–Carbon Bonds in Acetyl-CoA

The CO dehydrogenase (CODH) has a central role in methanogenesis from acetate and acts as a catalyst during acetyl-CoA. The CODHs are commonly found in anaerobes. Complete acetate molecules' oxidation to CO_2 is very common in anaerobes and this particular process helps in the reduction of different electron acceptors found in the cell. The acetyl-CoA cleavage by CODH gives carbonyl and methyl groups which are consequently oxidized to CO_2. In the energy-producing Ljungdahl-Wood process in homoacetogenic bacteria, CODH (acetyl-CoA synthase) is employed to catalyze acetyl-CoA synthesis by reducing CO_2. In methanogenic "Archaea" cells, carbon synthesis from CO_2 is also carried out by the CODH enzyme. In the pathway, one mole of CO_2 is decreased to methyl level in the CH_3-THF form. Another mole of CO_2 is decreased to the CO redox level which ultimately binds to the CODH enzyme. One of the corrinoid/Fe-S (corrinoid/iron-sulfur) takes up the methyl group of CH_3-THF. Then the methyl group is moved to the CODH enzyme. The CODH then catalyzes the production of acetyl-CoA from the CO, CoA moieties and CODH-bound methyl. The observation of increased CO-oxidizing activity in *M. barkeri* acetate-grown cells was the first indicator of involvement of CODH in methanogenesis from acetate (Krzycki et al., 1982).

3.4.2.3 Electron Transport and Bioenergetics during Acetate Activation

Conversion of acetate to CO_2 and CH_4 yields a small amount of energy given that acetate activation consumes an equivalent of one ATP.

$$CH_3COO^- + H^+ \rightarrow CH_4 + CO_2 (\Delta G^{0'} = -36 kJ/mol) \tag{3.10}$$

But over time these microorganisms have developed mechanisms that are very efficient at the conservation of energy. Though substrate-level phosphorylation is not so evident in the reactions in microorganisms, there exists evidence in support of chemiosmotic mechanism for ATP synthesis. A proton-motive force of -120mV is generated in the entire cells of *M. barkeri* which can degrade acetate. The dependency of electron transport on the membrane-bound proteins in the process of electron transfer from acetyl-CoA's carbonyl to "CoM-S-S-HTP" is majorly responsible for generating the above-mentioned proton-motive force. Three b-type cytochromes having low midpoint potentials ranging between -330mV to -182mV were discovered in *M. ethanosarcina, M. soehngenii*, and other acetate-grown cells (Kuhn et al., 1983). Although the CO_2-reducing pathway would not be able to proceed if coenzyme F_{420} would not act as an electron carrier (a 5-deazaflavin). But F_{420} does not act as an electron acceptor for CODH. However, F_{420} may perform reductive biosynthesis by providing electrons for the oxidation of the acetate methyl group to CO_2. Acetate-grown *M. barkeri* can also transfer F_{420} to F_{390} inside an ATP-dependent reaction similar to CO_2-reducing methanogenic species.

In *M. thermophila* a reconstitution of CODH complex, cytochrome linked hydrogenase, and ferredoxin has been done for a CO-oxidizing: H_2-developing system (Terlesky and Ferry, 1988). Though acetate-grown *M. thermophila* is unable to use H_2 -CO_2 for methanogenesis, they do exhibit H_2-dependent heterodisulfide reductase activity (Clements et al., 1993). Thus, it can be postulated that the CoM-S-S-HTP is decreased by the electrons provided by the oxidation of $H_{2.}$ Moreover, the membrane-bound electron carriers have the ability to couple acetate oxidation (carbonyl group) and H_2 evolution. According to this hypothesis, a possibility of synthesizing ATP by a chemiosmotic mechanism arises. In *M. barkeri* which are grown on acetate, there exists a coupling between CO oxidation to CO_2, $H_{2,}$ and proton translocation. The finding of methane production in *M. barkeri* by feeding on acetate has led to the suggestion that entire cells contain a membrane-bound electron transport mechanism that relies on H_2 to control electron flow when membranes are not present. Sodium has a significant role in the creation of methane from acetate and the sodium ion gradient generated happens simultaneously along with this process in *M. barkeri*. The transmembrane protein CH-H_4MSPT: HS-CoM methyltransferase, which was recently found in *M. barkeri* cultured on acetate (Fischer et al., 1992) catalyzes an exergonic process. This exergonic process is linked to a major electrogenic sodium extrusion in the CO_2 reduction and methanol utilization processes. However, the possibility of the reaction being combined with sodium extrusion in methanogenesis using acetate is yet to be explored and not much information is available about it.

3.5 PROTOCOL FOR BIOMETHANE PRODUCTION USING ISOLATES

Hungate technique for cultivation of the anaerobic methanogens was applied according to Balch et al. (1979).

Instruments: Anaerobic chamber, serum bottle, nitrogen gas supply and hydrogen gas supply, autoclave and pH meter.

Methodology

- The water in which the media will be prepared should be boiled at 100°C for 5 min. this step helps in expelling dissolved oxygen present in water.

TABLE 3.3
Media Composition for Cultivation of Acetoclastic Methanogens: DSM 287

K_2HPO_4	0.30 g
KH_2PO_4	0.30 g
$(NH4)_2SO_4$	0.30 g
NaCl	0.61 g
$CaCl_2$ x H_2O	0.14 g
$MgSO_4$ x 7 H_2O	0.13 g
Modified Wolin's mineral solution	10.00 ml
NH_4Cl	2.70 g
Na-acetate	2.50 g
Na-resazurin solution (0.1% w/v)	0.50 ml
$NaHCO_3$	5.00 g
Wolin's vitamin solution	10.00 ml
L-Cysteine-HCl x H_2O	0.30 g
Na_2S x $9H_2O$	0.30 g
Distilled water	1000.00 ml

Source: Stantscheff et al., 2014

- The boiled water is then cooled under N_2 gas bubbling so that no oxygen can dissolve back in it.
- The anaerobic water is then kept in anerobic gas chamber.
- All the media ingredients except bicarbonate, vitamins, cysteine and sulfide are mixed in anaerobic water (Table 3.3).
- It is further sparged with 80% H_2 and 20% CO_2 gas mixture for 30–45 min.
- Adjust the pH of the solution to 6.8 and dissolve bicarbonate salt in it.
- Disperse the media in serum bottles with 50% headspace and flush the headspace with 100% N_2.
- Close the bottle with butyl rubber septum and crimp it with aluminum cap.
- Autoclave the serum bottles and bring it back to room temperature.
- Add vitamin followed by cysteine and sulfide solution into the serum bottle by injecting it through butyl rubber cap using hypodermal syringe under anaerobic condition.
- Adjust the pH to 6.8 under anaerobic condition by using 1N HCl or 1N NaOH.
- For incubation use sterile 80% H_2 and 20% CO_2 gas mixture at one atmospheres of pressure.
- A 10 to 20% v/v inoculum is injected using syringe into the serum bottles.
- Inoculated serum bottles are then incubated at 37°C for 48 to 72 hours and checked for CH_4 production using Gas Chromatography (TCD detector).

Note: For cultivation of methylotrophic methanogens, the above-mentioned media can be modified, where acetate is replaced by methanol (5ml per 1000ml water). The purpose of the addition of Cysteine HCl and sodium sulfide in the methanogenic media is to scavenge any dissolved oxygen present in the media.

CONCLUSIONS

Methanogens are classified as acetoclastic, methyloclastic, and hydrogenotrophic methanogens based on their carbon source requirement. In a mixed microbial population, a synergy is observed between different types of methanogen and other organotrophic microorganisms. The *Methanobacterium, Methanosarcina, Methanospirillum Methanolobus, Methanococcus*, etc., are a few well-known methanogenic microorganisms. The complexity and uniqueness of metabolic pathways in different methanogens are the contrasting features of these microorganisms. Acetoclastic methanogenesis involves the activation of acetate in the form of acetyl-CoA. Activated acetate then cleaves to methane with the help of the CO dehydrogenase (CODH). Hydrogenotrophic methanogens produce methane by consuming CO_2 and H_2. The enzymes CHO-MFR, N-formylmethanofuran; CHO-H_4MPT and coenzyme M catalyzes the reaction between CO_2 and H_2. Almost all the methanogens are obligate anaerobes. Cultivating such microorganisms needs special skills such as the Hungate technique. The water in which the media will be prepared should be completely devoid of oxygen. Inoculation or the addition of vitamins is done under strictly anaerobic conditions. Understanding the fine details of obligate anaerobe cultivation could help in understanding the large-scale biomethane production system.

REFERENCES

Asakawa, Susumu, and Kazunari Nagaoka. "*Methanoculleus bourgensis, Methanoculleus olentangyi* and *Methanoculleus oldenburgensis* are subjective synonyms." *International journal of systematic and evolutionary microbiology* 53, no. 5 (2003): 1551–1552.

Balch, W. E., George E. Fox, Linda J. Magrum, Carl R. Woese, and RS281474 Wolfe. "Methanogens: reevaluation of a unique biological group." *Microbiological reviews* 43, no. 2 (1979): 260–296.

Barker, H. Albert. "On the biochemistry of the methane fermentation." *Archiv für Mikrobiologie* 7, no. 1 (1936): 404–419.

Bélaich, Jean-Pierre, Mireille Bruschi, and Jean-Louis García, eds. *Microbiology and Biochemistry of Strict Anaerobes Involved in Interspecies Hydrogen Transfer*. Vol. 54. Springer Science & Business Media, 2012.

Biavati, B., M. Vasta, and J. G. Ferry. "Isolation and characterization of *Methanosphaera cuniculi* sp. nov." *Applied and Environmental Microbiology* 54, no. 3 (1988): 768–771.

Blaut, M.-V., and Miller, Gottschalk, G. *Energetics of Methanogens in the Bacteria*, Vol. XII, Sokatch, J.-R., Ornston, L.-N. (eds.) Academic Press, San Diego (1990):505–537.

Bleicher, K., G. Zellner, and J. Winter. "Growth of methanogens on cyclopentanol/CO2 and specificity of alcohol dehydrogenase." *FEMS microbiology letters* 59, no. 3 (1989): 307–312.

Boone, David R., Richard L. Johnson, and Yitai Liu. "Diffusion of the interspecies electron carriers H2 and formate in methanogenic ecosystems and its implications in the measurement of K_m for H_2 or formate uptake." *Applied and Environmental Microbiology* 55, no. 7 (1989): 1735–1741.

Bryant, Marvin P., and David R. Boone. "Emended description of strain MST (DSM 800T), the type strain of Methanosarcina barkeri." *International Journal of Systematic and Evolutionary Microbiology* 37, no. 2 (1987): 169–170.

Burggraf, S., K. O. Stetter, P. Rouviere, and C. R. Woese. "*Methanopyrus kandleri*: an archaeal methanogen unrelated to all other known methanogens." *Systematic and Applied Microbiology* 14, no. 4 (1991): 346–351.

Clements, Andrew P., Robert H. White, and James G. Ferry. "Structural characterization and physiological function of component B from *Methanosarcina thermophila*." *Archives of Microbiology* 159, no. 3 (1993): 296–300.

Conrad, Ralf, T. J. Phelps, and J. G. Zeikus. "Gas metabolism evidence in support of the juxtaposition of hydrogen-producing and methanogenic bacteria in sewage sludge and lake sediments." *Applied and Environmental Microbiology* 50, no. 3 (1985): 595–601.

Cord-Ruwisch, Ralf, Hans-Jürgen Seitz, and Ralf Conrad. "The capacity of hydrogenotrophic anaerobic bacteria to compete for traces of hydrogen depends on the redox potential of the terminal electron acceptor." *Archives of Microbiology* 149, no. 4 (1988): 350–357.

Fischer, R., P. Gärtner, A. Yeliseev, and R. K. Thauer. "N 5-Methyltetrahydromethanopterin: coenzyme M methyltransferase in methanogenic archaebacteria is a membrane protein." *Archives of Microbiology* 158, no. 3 (1992): 208–217.

Huber, Harald, Michael Thomm, Helmut König, Gesa Thies, and Karl O. Stetter. "*Methanococcus thermolithotrophicus*, a novel thermophilic lithotrophic methanogen." *Archives of Microbiology* 132, no. 1 (1982): 47–50.

Huser, Beat A., Karl Wuhrmann, and Alexander JB Zehnder. "*Methanothrix soehngenii* gen. nov. sp. nov., a new acetotrophic non-hydrogen-oxidizing methane bacterium." *Archives of Microbiology* 132, no. 1 (1982): 1–9.

Jetten, Mike SM, Alfons JM Stams, and Alexander JB Zehnder. "Acetate threshold values and acetate activating enzymes in methanogenic bacteria." *FEMS Microbiology Ecology* 6, no. 4 (1990): 339–344.

König, Helmut, and Karl O. Stetter. "Isolation and characterization of Methanolobus tindarius, sp. nov., a coccoid methanogen growing only on methanol and methylamines." *Zentralblatt für Bakteriologie Mikrobiologie und Hygiene: I. Abt. Originale C: Allgemeine, angewandte und ökologische Mikrobiologie* 3, no. 4 (1982): 478–490.

Kristjansson, Jakob K., Peter Schönheit, and Rudolf K. Thauer. "Different Ks values for hydrogen of methanogenic bacteria and sulfate reducing bacteria: an explanation for the apparent inhibition of methanogenesis by sulfate." *Archives of Microbiology* 131, no. 3 (1982): 278–282.

Krzycki, Joseph Adrian, R. H. Wolkin, and J. G. Zeikus. "Comparison of unitrophic and mixotrophic substrate metabolism by an acetate-adapted strain of *Methanosarcina barkeri*." *Journal of Bacteriology* 149, no. 1 (1982): 247–254.

Kühn, Wolfgang, Klaus Fiebig, Hans Hippe, Robert A. Mah, Beat A. Huser, and Gerhard Gottschalk. "Distribution of cytochromes in methanogenic bacteria." *FEMS Microbiology Letters* 20, no. 3 (1983): 407–410.

Kurr, Margit, Robert Huber, Helmut König, Holger W. Jannasch, Hans Fricke, Antonio Trincone, Jakob K. Kristjansson, and Karl O. Stetter. "*Methanopyrus kandleri*, gen. and sp. nov. represents a novel group of hyperthermophilic methanogens, growing at 110 C." *Archives of Microbiology* 156, no. 4 (1991): 239–247.

Lauerer, Gerta, Jakob K. Kristjansson, Thomas A. Langworthy, Helmut König, and Karl O. Stetter. "*Methanothermus sociabilis* sp. nov., a second species within the Methanothermaceae growing at 97 C." *Systematic and Applied Microbiology* 8, no. 1–2 (1986): 100–105.

Lundie, L. L., and J. G. Ferry. "Activation of acetate by *Methanosarcina thermophila*: purification and characterization of phosphotransacetylase." *Journal of Biological Chemistry* 264, no. 31 (1989): 18392–18396.

Mah, Robert A., and Daisy A. Kuhn. "Transfer of the type species of the genus *Methanococcus* to the genus Methanosarcina, naming it *Methanosarcina mazei* (Barker 1936) comb. nov. et emend. and conservation of the genus Methanococcus (Approved Lists 1980) with *Methanococcus vannielii* (Approved Lists 1980) as the type species: Request for an opinion." *International Journal of Systematic and Evolutionary Microbiology* 34, no. 2 (1984): 263–265.

McInerney, M. J., M. P. Bryant, R. B. Hespell, and J. W. Costerton. "*Syntrophomonas wolfei* gen. nov. sp. nov., an anaerobic, syntrophic, fatty acid-oxidizing bacterium." *Applied and Environmental Microbiology* 41, no. 4 (1981): 1029–1039.

McInerney, Michael J., and P. Shawn Beaty. "Anaerobic community structure from a nonequilibrium thermodynamic perspective." *Canadian Journal of Microbiology* 34, no. 4 (1988): 487–493.

Miller, Terry L., and Meyer J. Wolin. "*Methanosphaera stadtmaniae* gen. nov., sp. nov.: a species that forms methane by reducing methanol with hydrogen." *Archives of Microbiology* 141, no. 2 (1985): 116–122.

Pine, Martin J., and Wolf Vishniac. "The methane fermentations of acetate and methanol." *Journal of Bacteriology* 73, no. 6 (1957): 736–742.

Platen, Harald, and Bernhard Schink. "Methanogenic degradation of acetone by an enrichment culture." *Archives of Microbiology* 149, no. 2 (1987): 136–141.

Raskin, Lutgarde, Julie M. Stromley, Bruce E. Rittmann, and David A. Stahl. "Group-specific 16S rRNA hybridization probes to describe natural communities of methanogens." *Applied and Environmental Microbiology* 60, no. 4 (1994): 1232–1240.

Robinson, Joseph A., and James M. Tiedje. "Competition between sulfate-reducing and methanogenic bacteria for H_2 under resting and growing conditions." *Archives of Microbiology* 137, no. 1 (1984): 26–32.

Robinson, Joseph A., and James M. Tiedje. "Kinetics of hydrogen consumption by rumen fluid, anaerobic digestor sludge, and sediment." *Applied and Environmental Microbiology* 44, no. 6 (1982): 1374–1384.

Rouviere, Pierre E., and R. S. Wolfe. "Novel biochemistry of methanogenesis." *Journal of Biological Chemistry* 263, no. 17 (1988): 7913–7916.

Stantscheff, R., J. Kuever, A. Rabenstein, K. Seyfarth, S. Dröge, and H. König. "Isolation and differentiation of methanogenic Archaea from mesophilic corn-fed on-farm biogas plants with special emphasis on the genus *Methanobacterium*." *Applied Microbiology and Biotechnology* 98, no. 12 (2014): 5719–5735.

Terlesky, K. C., and J. G. Ferry. "Ferredoxin requirement for electron transport from the carbon monoxide dehydrogenase complex to a membrane-bound hydrogenase in acetate-grown *Methanosarcina thermophila.*" *Journal of Biological Chemistry* 263, no. 9 (1988): 4075–4079.

Thauer, R. K., and J. G. Morris. "Metabolism of chemotrophic anaerobes: old views and new aspects." In *Symposia of the Society for General Microbiology (Cambridge) [SYMP. SOC. GEN. MICROBIOL. (CAMB.).]. 1984*. 1984.

Woese, Carl R., Otto Kandler, and Mark L. Wheelis. "Towards a natural system of organisms: proposal for the domains Archaea, Bacteria, and Eucarya." *Proceedings of the National Academy of Sciences* 87, no. 12 (1990): 4576–4579.

Zinder, S. H., S. C. Cardwell, T. Anguish, M. Lee, and M. Koch. "Methanogenesis in a thermophilic (58 C) anaerobic digestor: *Methanothrix* sp. as an important aceticlastic methanogen." *Applied and Environmental Microbiology* 47, no. 4 (1984): 796–807.

4 Bioethanol Production Process

4.1 INTRODUCTION

In the earlier twentieth century, Henry Ford, the founder of the innovative vehicles, designed his Model T intending to fuel it with ethanol derived from cereals. By 1938, factories in Kansas were generating 18 million gallons of ethanol/year (approx. 54,000 t/year), demonstrating Ford's commitment to the usage of this fuel (Di Nicola et al., 2011). However, because of the abundant supply of oil and natural gas after WWII, interest in ethanol started to decline. Following the first oil crisis at the end of the 1970s, many oil firms started selling gasohol, gasoline that contained 10% ethanol, to take advantage of the tax savings available for ethanol. However, since it already had rivals in the automobile market, including MTBE (methyl tert-butyl ether), which was superior to ETBE (ethyl tert-butyl ether) in terms of economic as well as efficiency, bioethanol did not instantly achieve the popularity it deserved. Following the discovery that MTBE was highly polluting, it was prohibited, and bioethanol reemerged as one of the most promising potential options to lowering emission of CO_2. Another element that aided the relaunch of bioethanol was a rising understanding that we are approaching the so-called turning point or the crucial point of no return when the demand curve for oil crosses the diminishing curve of its supply. However, there is an ethical dilemma that is especially relevant to bioethanol, but which also impacts other biofuels. Biofuels are mostly made from raw materials like grains and plants that may otherwise be used in the food sector. To address this issue, current research has focused on *Miscanthus giganteus*, an inedible perennial herbaceous plant contained calorific content of roughly 4200 kcal/kg of dry matter. The ethical issues might be solved by using lignocellulose materials, the annual wheat waste, or municipal solid waste (about 5%, which may generate roughly 9.3 Gl of bioethanol).

4.1.1 Usage of Ethanol as Fuel

Bioethanol may be utilized in a variety of ways, including in diesel engines add 5 to 10% diesel oil, mixing 10–85% with gasoline for internal combustion engines, or replacing 0–100% of the gasoline in FFV (flexible fuel vehicles). The FFVs

DOI: 10.1201/9781003224587-4

number on the road is steadily rising: sales in Brazil have reached 400,000/year, and over 1.5 million of them (mostly public automobiles) on the road in the United States; in Europe, Sweden has about 15,000 of these automobiles that run on E85 (85% ethanol). Improved bioethanol-fuel engines and fuel cells that employ internal bioethanol reforming to generate hydrogen are also being researched.

4.1.1.1 Ethyl Tert-butyl Ether (ETBE)

The ETBE is a bioethanol product with high-octane derived primarily by heating ethanol and reacting it with isobutylene (an oil refining byproduct) using different catalysts. As a result, it is classified as a partly renewable resource. It improves upon the functional and technological characteristics of the alcohol from which it is derived. It also lacks the latter's concerns with miscibility and volatility with petrol, as well as a high-octane rating. Because it is ether, it has oxygen in molecule, which allows it to aid in the reduction of pollution emissions from the vehicle. Anti-detonating factors and RVP ("Reid vapor pressure") were studied by Da Silva et al. in 2005, and they found that adding ETBE increased anti-detonating components and decreased RVP without interfering with the volatility required for initiation of a cold engine. The ETBE also increased RVP. A bioethanol-based ETBE provides all of the similar advantages as bioethanol, including lower pollutant emissions and a high-octane number, but without logistical and technical issues that come with bioethanol's alcoholic nature. BioETBE also aids the expansion of biofuels in the transportation industry.

4.1.1.2 Diesel and Bioethanol Mixtures (E-diesel)

To meet European Union Directive 2003/30/CE, European nations have developed and expanded their usage of bioethanol and diesel mixes in diesel engines (by 2010, biofuels must comprise at least 5.75% of the global fuel supply), and increased need for diesel cars necessitates the disposal of a surplus of petrol in refineries. Low lubrication issues, lesser injection capacity, and increased volatility all contribute to the drawbacks of so-called "e-diesel" (resulting in rise in unburned hydrocarbon emissions), as well as lower miscibility (Lapuerta, 2007). Lapuerta et al., in particular, investigated several diesel-bioethanol combinations under various temperature conditions, additive and water content, mapping out the zones of stability and kinetic separation in the mixes and drawing conclusions such as:

- The presence of water aids in the ethanol phase separation.

  The combination gets more stable as the temperature rises, and the ethanol solubility in diesel also rises.

  The greater the temperature of the mixture, the more sensitive it is to additives and water content effects.

  In places where winter temperatures seldom fall below -5°C, bioethanol mixes with up to 10% bioethanol concentration (v/v) may be utilized in diesel engines.
- To avoid phase separation, the inclusion of additives that improve stability may allow for a wider range of ethanol concentrations in mixes and a broader geographic distribution of their use.

4.2 FEEDSTOCK FOR BIOETHANOL PRODUCTION

In comparison to traditional biofuels, biofuels made from wood cellulose and other organic resources have some distinct benefits. Burning of cellulose-derived ethanol reduces emissions by 87% compared to burning petrol, but cereal-derived ethanol only reduces emissions by 28%. It takes 16 times as much energy to create ethanol from cellulose, five times as much to make petrol, and 1.3 times as much to make maize ethanol (Martinez, 2009). The issue is how to break the molecular bonds to get fermentable sugars from it? Farm as well as forest trash, scrap wood, grasses are grown-up for energy uses or even municipal sewage waste (MSW) may be used as initial material for biofuels production. Lignocellulose is found in the cell walls of vegetables and is composed of cellulose microfibers found in lignin, pectin, and hemicellulose (Table 4.1). Ethanol is produced by fermenting monomeric sugars, which have been depolymerized to create monomeric sugars. This biomass is made up mostly of three basic polymers, which are cellulose, lignin, hemicellulose (for example, xylose), and minor elements (acids, essential oils, minerals, and salts).

There are various approaches for producing energy and biofuels from lignocellulosic biomass (LCB). The LCB can be categorized agricultural residues, forest residues, energy crops, municipal organic wastes (Figure 4.1). Biochemical procedures are usually carried out using LCBs with a C/N (carbon/nitrogen) ratio below 30 and humidity at the collection over 30%. Enzymes, microorganisms, and mushrooms are used to catalyze chemical reactions in these procedures. When the LCB supplied has a C/N ratio over 30 and a humidity content of less than 30%, thermochemical techniques may be applied. Novel biofuels including butanol, bio-H_2, gamma-valerolactone, and dimethylfuran have been produced from LCB in recent years (Ortiz et al., 2008). The generation of biofuels from LCB still has certain engineering challenges, but researchers are working hard to solve them. A long-term goal of LCB's development is to minimize deprivation while providing structural toughness as well as hydrolytic stability in the cell walls of plants. Polysaccharides and lignin are linked together through ester and ether linkages to provide this resilience. The LCB, on the other hand, is seen as the most potential raw material to generate renewable biofuels due to its accessibility, lower price, as well environmental friendliness.

TABLE 4.1
Composition of Lignocellulosic Biomass

Feedstock	Cellulose (%)	Hemicellulose	Lignin
Energy crops	21–54	5–30	5–10
Grasses	25–40	25–50	10–30
Softwoods	45–50	25–35	25–35
Hardwoods	45–55	24–40	18–25

Source: Gírio et al., 2010

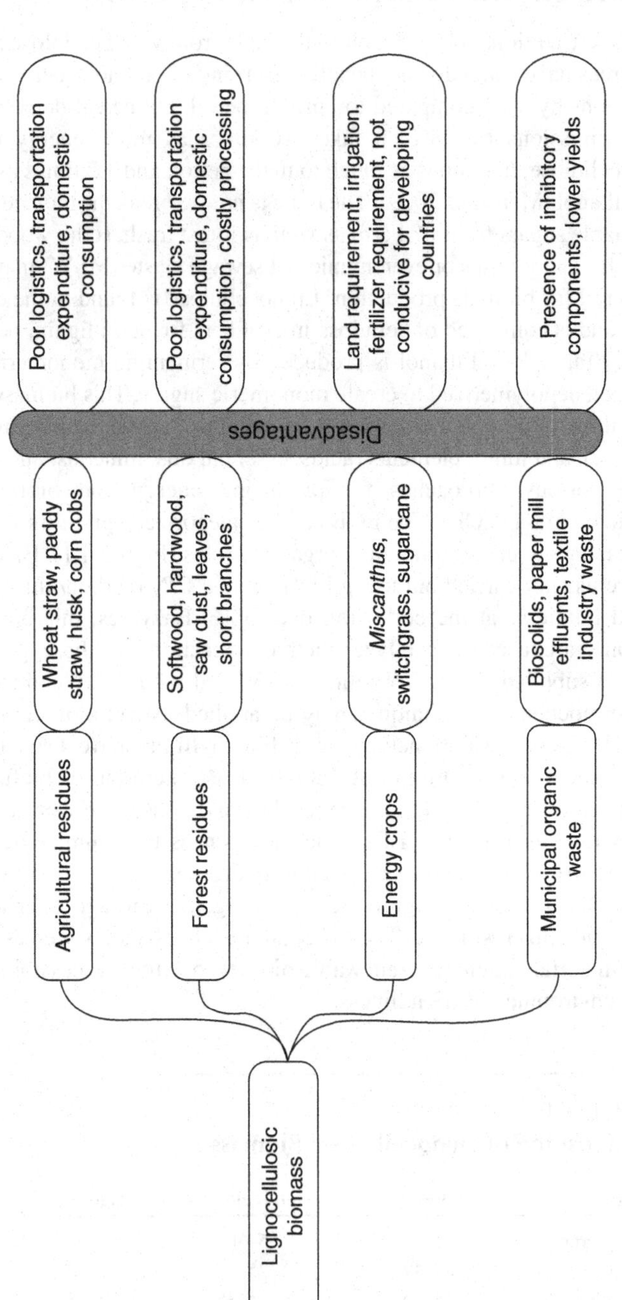

FIGURE 4.1 Potential lignocellulosic biomass and their disadvantages as feedstock.

4.2.1 Pre-treatments of Lignocellulosic Biomass

Microscopic and macroscopic raw materials, as well as their chemical compositions, may be altered via the application of these mechanisms. The hemicellulose is solubilized, the crystallinity is reduced, and the accessible surface area and porosity of the substrate are increased as a result. The following criteria must be met for pre-treatment to be effective: It must promote sugar production or assist future sugar creation during hydrolysis while eliminating any deterioration or carbohydrates loss and preventing the generation of byproducts that might impede later fermentation and hydrolysis processes, all at an affordable price (Balat, 2009). Pre-treatments, such as chemical, physical, biological, AFEX ("ammonia fiber explosion"), and steam explosion are very important before enzymatic hydrolysis.

Mechanical or non-mechanical physical pre-treatments are possible. Grinding and milling are two mechanical physical pre-treatments that lower the substrate while simultaneously increasing the surface area to volume ratio, facilitating the hydrolysis of cellulose more easily. To lessen the crystallinity of the cellulose, "ball milling" might be employed, but this method is both costly and time-consuming (it takes about a week to finish), making it impractical for commercial application. Internal and external pressures are used in combination to decompose the lignocellulose in non-mechanical pre-treatments.

To lower the proportion of crystalline cellulose, chemical pre-treatments are often used. However, using this form of pre-treatment causes issues for plants because the chemical agents' severe operating conditions require that all structural components be able to withstand them. One of the most popular chemical pre-treatments is an alkaline treatment, which delignifies and solubilizes the glycan, and another is a NaOH treatment, which dissolves the lignocellulose biomass and destroys the lignin structure of the biomass. It is also essential to dilute sulfuric acid prior to hydrolysis, however the hydrolyzed end products are rarely fermentable when paired with diluted acid hydrolysis. In addition to pre-treatment with H_2O_2 (hydrogen peroxide), which utilizes oxidative delignification for solubilizing and separating lignin and dissolve lignocellulose matrices, other methods include pre-treatment with ozone degrading lignin polymers, and pre-treatment with LHW (liquid hot water), which is typically used on alfalfa. This approach is as efficient as diluted acid hydrolysis under ideal circumstances (Laser et al., 2002), without the requirement for any acid or neutralization products.

Enzymes, which are already beneficial in industry applications on wood waste, are used in the processing of trash and pulp in biological pre-treatments. Basidiomycetes such as *Pleurotus ostreatus* generate enzymes that have been studied extensively in the past: these enzymes are homologous proteins with distinct properties depending on which phenols are replaced. The *Phanerochaete chrysosporium* is another basidiomycetes fungus that may help with delignification (Palmieri et al., 1997). Decomposition of hemicellulose, cellulose and lignin fibers and hydrolysis of biomass are achieved by using saturated steam at very high pressure and temperature. A high-pressure steam stream is applied into a sealed chamber that contains lignocellulose, the pressure is reduced, causing the matrix and steam to expand, and then explosive decompression is obtained through an orifice, causing the substrate's cellular structure to be disrupted and the acetyl groups in the hemicellulose to be decomposed. In certain circumstances (e.g., angiosperm), using acid catalysts like SO_2 or H_2SO_2 for making easier access of cellulose-rich components to enzymes is desirable. Although SO_2 gas is more effective in attacking fibers (Shevchenko, 2000), its usage requires careful consideration of the operating conditions where steam explosion occurs. In reality, the optimum balance between strong enzymatic hydrolysis (attainable only under extreme conditions) and high recovery of hemicellulose-containing components in the monomeric sugars form (which requires comparatively fewer extreme conditions) becomes important (Silverstein, 2009). Consequently, a severity indicator (Overend and Chornet, 1987) has been designed, which compares pre-treatment timing and temperatures, and assumes that this process follows Arrhenius equation and has 1st order kinetics. The R_0 indicates:

$$R_o = t\exp\left[\frac{T_r - T_b}{14.75}\right] \tag{4.1}$$

where the pre-treatment time in minutes is denoted by 't,' the reaction temperature in degrees Celsius by 'T_r,' baseline temperature in (100°C) by 'T_b', and the conventional activation energy is 14.75 which assumes the first-order conversion. If the sulfuric acid version is utilized, the severity parameter "M_0" is slightly changed as follows:

$$M_o = C^n t\exp\left[\frac{T_r - T_b}{14.75}\right] \tag{4.2}$$

where chemical concentration is denoted by C (in wt%) and arbitrary constant by n (Chum, 1991). Pre-treatment with liquid ammonia and steam explosion is called AFEX: an NH_3 solution of 1–2 kg NH_3/kg of dry biomass is put under the pressure the earlier humidified lignocellulose content. This process works well with non-woody materials like newspaper and bagasse, but not so well with wooden "soft" material. Not directly, but it makes it easier for enzymes to break polymers down into their constituent parts (cellulose and hemicellulose). Because carbon dioxide is less expensive than ammonia and due to the presence of lignin-derived contaminants in the alcohol waste product, ammonia may be substituted with carbon dioxide. The AFEX and LHW are the most effective pre-treatments for agricultural waste, whereas steam pre-treatment yields a high sugar output from both forests as well as agricultural waste.

4.2.2 The Hydrolysis Process

The following law governs hydrolysis:

$$(C_6H_{10}O_5)_n + nH_2O \rightarrow nC_6H_{12}O_6 \tag{4.3}$$

and there are mostly two types: enzymatic or acid (using concentrated/diluted acids). Because it includes components other than glucose, including lignin and hemicellulose, hydrolyzing lignocellulose biomass is more complex than hydrolyzing pure cellulose. Acid hydrolysis of lignocellulose biomass generates mostly xylose, whereas the cellulose and lignin components remain unaltered. To hydrolyze xylan, moderate acidity is required, but cellulose needs more extreme conditions because of its crystallized structure. To depolymerize the hemicellulose, 1% sulfuric acid must be used in the hydrolysis process. In most cases, there are two steps to this procedure. The rapidity of the processes engaged in this form of hydrolysis is one of the most significant properties since it allows the process to continue. To accelerate the acid distribution, the raw material is broken mechanically down into small particles of a few millimeters in diameter. In contrast, hydrolysis with concentrated acids (10 to 30%), converts cellulose to glucose and hemicellulose to xylose quickly and completely, with minor degradation. Sulfuric and hydrochloric acids, as well as hydrogen fluoride, are the most often utilized acids.

Using this acid hydrolysis method, sugars are recovered with high efficiency (about 90% of cellulose and hemicellulose are depolymerized into monomeric sugars). Compared to the diluted acid solution, this approach reduces manufacturing costs, particularly if acids were extracted as well as re-concentrated. Sugars and acids in mixture are separated after ion exchanging, multiple-effect evaporators are used to re-concentrate the acid. Pellets generated from the residual solid fractions, which include lignin, may be used as a source of energy.

There are two steps to the process of concentrated acid hydrolysis: a concentrated acid (70%) breaks the hydrogen bonds among cellulose chains in the initial stage of the process; One isolated cellulose chain undergoes a hydrolytic process in the second step of hydrolysis. Many structural factors of the substrate, including the quantity of cellulose in the substrate, the surface area, and the crystallinity of cellulose, slow down the enzymatic hydrolysis of lignocellulose materials. To make the biomass more vulnerable to hydrolysis, pre-treatments are required. A cocktail of enzymes that can break polymeric chain links is also required for this reason. The xylanase, cellulase, mannoxidase, and hemicellulase are among the hydrolytic enzymes often used in this cocktail. Enzymatic cellulose degradation is a difficult process to understand since it occurs at the interface of the liquid and solid phases, with enzymes as the mobile components. In general, degradation is divided into two phases: a rapid first phase and a second slower phase which might persist until all of the substrates have been consumed. The accessible part of cellulose hydrolyzes quickly, then the gradual ingested enzyme molecules activation which is thought to be the source of this behavior.

Chopping the biomass improves the surface area available for enzymes as well as lowers the cellulose crystallinity and polymerization, allowing for the use of fewer enzymes and lower manufacturing costs. For lignocellulose materials hydrolysis, both fungi and bacteria are capable of producing cellulose. Bacteria may be thermophylic or mesophylic, anaerobic or aerobic. *Cellulomonas*, *Clostridium*, *Thermomonospora*, *Bacillus*, *Bacteroides*, *Ruminococcus*, *Acetovibrio*, *Erwinia*,

Streptomyces, and *Microbispora* are among the most often utilized bacteria. Depending on where an enzyme performs its reaction, it may either be classed as extracellular or intracellular (or cell-associated). To induce changes in the cell mass components, the primary role of extracellular enzymes is to transform the substrate into a medium for external use. Intracellular enzymes, on the other hand, are unable to transform a substrate until it has spread throughout the cellular mass.

The synergic activity of the enzyme's endoglucanase (EG or endo-1, 4-β-glucanase), exoglucanase (CBH or cellobiohydrolase), and α-glucosidase is the most commonly recognized method for the enzymatic cellulose hydrolysis. α-glucosidase is intracellular whereas CBH and EG both are extracellular enzymes. EG causes the cellulose chains to be randomly disrupted, resulting in significant degradation. With α-1, 4- glucoside bonds hydrolyzing, it works to form new chains. Exoglucanase breaks apart the chain ends, allowing solubilize cellobiose and glucose to be released. BGL catalyzes the hydrolysis of cellobiose into glucose, therefore removing the inhibiting cellobiose; subsequently, the hydrolysis of cellobiose into glucose is completed by BGL, which acts as a catalyst, so completing the metabolic pathway. Microorganisms including filamentous fungus, like Trichoderma sp., generate most cellulase and hemicellulase, which may be employed in their native state or modified genetically (*Trichoderma reesei*, *Trichoderma viride*, *Trichoderma longibrachiatum*). The *Trichoderma reesei*'s primary enzymes are CBH I and II, whereas the dominating endoglucanases are EG I and II. Temperature (a 20 to 30°C rise in temperature results in a 3 to 5 times rise in end products) is one of the factors that influence enzyme activity. The most important aspect of temperature is the possibility of unintended denaturation if the temperature is extremely high (Balat, 2008). Catalytic or non-catalytic enzymatic hydrolysis has consistently produced large yields of glucose (>90%) and xylose (>80%).

4.2.3 Fermentation Process

After hydrolysis, microorganisms, especially yeasts, will ferment the hydrolyzed compounds (Hahn-Hägerdal et al., 2006). Because the hydrolysis products mostly comprise of xylose, glucose, cellobiose, and arabinose, the microbes utilized to ferment them should be able to ferment them all effectively to produce ethanol on a wide scale. The following are the reactions that occur when xylose and glucose are combined:

$$3C_5H_{10}O_5 \rightarrow 5C_2H_5OH + 5CO_2 \tag{4.4}$$

$$C_6H_{12}O_6 \rightarrow 2C_2H_5OH + 2CO_2 \tag{4.5}$$

SHF ("Separate hydrolysis and fermentation"), where two processes are performed in separate units, is the traditional technique for fermenting hydrolyzed biomass. SSF, or simultaneous saccharification and fermentation, is a typical option that combines fermentation as well as hydrolysis in the single unit. Finally, there is CBP (consolidated bioprocessing) as the last alternative.

Saccharification is the hydrolysis of the lignocellulose solid fraction that occurs during the SHF process. Liquids are fermented and distilled for bioethanol, retaining only the unfermented xylose, which is fermented once more in a different reactor and finally subjected to a final distillation phase. The fundamental benefit of this method is that it separates the hydrolysis and fermentation mechanisms, allowing for the adoption of ideal operating conditions for each. The enzymes are unrestricted in their ability to act at extreme temperatures, whereas the microorganisms may initiate fermentation at lower temperature. The enzymatic hydrolytic process been used to produce ethanol from a mix of municipal solid waste: in this case, enzyme recycling was improved using micro- and ultra-filtering procedures, thus achieving the hydrolysis of 90% of the cellulose with a net enzyme load of 10 FPU g^{-1} of cellulose (where FPU stands for filter paper unit) (Sánchez & Cardona, 2008).

Enzymatic fermentation as well as hydrolysis take place concurrently during the SSF procedure. Bacterial cells easily transform glucose produced by hydrolysis of cellulose into ethanol since the two processes are mutually reinforcing. Because yeast fermentation reduces the inhibitory influence of glucose and cellobiose end products, SSF generates the most bioethanol at the lowest cost; resulting in a decreased requirement for enzymes. Heterogeneous natural materials comprising complex polymers including lignin, lignocellulose, and pectin are used in this type of procedure. With SSF, the major benefits are a quicker hydrolysis rate attributable to the breakdown of sugars that impede cellulase activity; low enzyme requirement; high yielding product; fewer sterile conditions; shorter processing times; lower reactor dimensions; and shorter processing time (Sun and Cheng , 2002). There are some drawbacks to this method, the most prominent of which is the fact that the fermentation and hydrolysis processes must be completed in suboptimum conditions. For such method to be successful, the selection of microorganisms and preparation is very essential. The enzyme cocktail used to hydrolyze the cellulose must also be stable.

SSF's normal operating conditions for *Saccharomyces cerevisiae* cultures include a pH of 4.5 and temperatures approximately 310 K. Simultaneous saccharification and co-fermentation (SSCF), a new form of this technique where 5- and 6-carbon sugars to be fermented concurrently, has recently been tested. Hydrolysis constantly releases hexose sugars in SSCF, escalating the glycolysis rate and enabling the pentose sugars to ferment quickly and provide an optimum output.

A single CBP process involves four types of biologically mediated conversions: the glycolytic enzymes production (cellulase and hemicellulase), hydrolysis of the pretreated biomass carbohydrate component for extracting sugars, six-carbon sugars (mannose, glucose, and galactose), whereas five-carbon sugars (arabinose and xylose).

The primary difference between CBP and other procedures is that there is no specific step for the cellulose synthesis. CBP, also known as DMC ("Direct microbial conversion"), needs only one microbial community for both cellulose synthesis and fermentation. Finding an organism that can concurrently make cellulose and ethanol with an optimal output is the main drawback of this strategy. The *Clostridium thermocellum* was used for enzyme synthesis, glucose fermentation, and cellulose hydrolysis, whereas *Clostridium thermosaccharolyticum* enables concurrent pentose sugars conversion which is derived from hemicellulose hydrolysis into ethanol as described by Wyman (1994).

The *Clostridium thermocellum* increases substrate conversion by 31% compared to *Saccharomyces cerevisiae* or *Trichoderma reesei*. Recent research has concentrated on cellulase generation and high ethanol yields employing strains of *Escherichia coli*, *Zymomonas mobilis*, *Klebsiella oxytoca*, and the yeast *Saccharomyces cerevisiae*. Microcrystalline cellulose hydrolysis yield was enhanced and anaerobic growth in amorphous cellulose was facilitated by expressing cellulase in *Klebsiella oxytoca*. In *Saccharomyces cerevisiae*, many cellobiohydrolases have been functionally expressed. New stable species of microbes able of processing cellulose biomass into bioethanol will be formed as a result of genetic engineering and metabolic investigations, resulting in advancements in the commercial production process of bioethanol (Lynd, 2006).

Microorganisms utilized in the fermentation activity should be able to work well with both polysaccharide as well as monosaccharide sugars, making them versatile. The survival of such microorganisms is dependent on maintaining a constant pH, and the most of bacteria cannot withstand bioethanol levels of more than 10–15% (w/v). The most often utilized microorganisms is *Saccharomyces cerevisiae* as it produces a significant amount of ethanol from hexose sugars and tolerates bioethanol and inhibiting chemicals effectively. The inability to absorb C6 sugars is a significant disadvantage.

The *Klebsiella oxytoca*, *Escherichia coli*, and *Zymomonas mobilis* seem to be the most effective ethanol-producing bacteria for commercial use. The *Zymomonas* sp., in particular, has shown an ability to promptly and effectively generate bioethanol from glucose-based raw materials, with a five times greater output in comparison to other yeasts. The ethanol it generates during fermentation of glucose has a theoretical yield of 97% and may be found in ratios up to 12% (w/v). These bacteria can effectively produce bioethanol from saccharose and fructose (C5) sugars, but not from C6 sugars.

The *Pichia stipitis*, *Candida parapsilopis*, and *Candida shehatae* are examples of yeasts that naturally ferment xylose, and they do so by using XR (xylose reductase) to transform xylose to xylitol, followed by XDH (xylitol dehydrogenase) for transforming xylitol to xylulose. The heterologues XDH and XR of *Pichia stipitis*, as well as XK (Xylulose Kinase) of *Saccharomyces cerevisiae*, may be used to produce bioethanol from xylose in recombinant *Saccharomyces cerevisiae*. Table 4.2 lists the most widely utilized microorganisms and bacteria (Balat et al., 2008), as well as the main metrics used to evaluate the effectiveness of different kinds of fermentation.

Fermentation may take place in a variety of methods, including continuously, discontinuously, with immobilized cells, and batch-fed (Chandel et al., 2009).

The development of inhibitors is a concern with enzymatic hydrolysis. Some concentrations of glucose, cellobiose, or pre-treatment products including furfural and organic acids have a significant impact on the activity of enzymes. Depending on the circumstances under which enzymatic hydrolysis occurs, inhibitors form.

To get the optimum solubilization and recovery of hemicellulose (low-severity), the best hydrolysis of the water-insoluble cellulosic component (high-severity), and compromising these two conditions (medium severity), it is possible to choose the conditions. R0 (severity factor) is linked to the ambient pH, and the combined severity (CS) index represents the severity of the earlier discussed variables.

Its value is calculated as follows:

$$CS = \log R_0 \text{-pH} \tag{4.6}$$

Furfural and hydroxymethylfurfural (HMF) are formed when the fermentable sugars concentration drops below the level that yields the maximum levels of glucose as well as mannose, and these acids disintegrate into formic and levulinic acids.

A CS of roughly '3' is recommended for optimal fermentability and a high output from fermentable sugars (Palmqvist and Hahn-Hägerdal, 2000). Carbohydrate degradation, equipment, wood extracts, lignin decomposition, and their decomposition are all examples of potential inhibitors. Organic, furanes, acid, and phenolic substances may all be grouped based on their molecular structure. Furane derivatives, including furfural and 5-HMF (5-hydroxymethylfurfural),

TABLE 4.2
Commonly Used Bacteria and Microorganisms

Species	Characteristics
Clostridium thermocellum	The capability of converting cellulose to acetic acid and ethanol directly.
Clostridium acetobutilicum	Useful in the butanol and acetone fermentation of xylose.
Escherichia coli	Xylose is fermented by native strains into a Bioethanol mixture.
Klebsiella oxytoca	Cellobiose and xylose are rapidly fermented by native strains.
Latobacillus casei	Lactose ferments well, making it especially suitable for whey bioconversion.
Lactobacillus pentoaceticus	Consumes arabinose and xylose.
Lactobacillus pentosus	Homolactic fermentation: Sulfite waste liquors are used by certain strains to generate lactic acid.
Lactobacillus xylosus	If nutrients are available, cellobiose is used: glucose, L-arabinose, and D-xylose are used.
Zymomonas mobilis	Normally, fructose and glucose are fermented.
Lactobacillus plantarum	Cellobiose is consumed faster than arabinose, glucose, or xylose.

Source: Balat et al., 2008

aliphatic acids, like levulinic acid, formic acid, and acetic acid, as well as phenolic chemicals, in particular, are fermentation inhibitors.

Certain polymeric materials may be formed from the furane derivatives. Process water recirculation is important because of the development of inhibitory chemicals in the production process. Non-concentrated hydrolyzed products have a modest inhibitory effect on *Saccharomyces cerevisiae*, whereas 5-times higher quantity of nonvolatile elements almost entirely suppress ethanol production by *Saccharomyces cerevisiae* (Palmqvist et al., 1996). Toxic chemicals are formed when enzymes are inhibited, which has a deleterious effect on both enzymatic fermentation and hydrolysis. Pre-treatments such as steam explosions and hydrolysis due to low concentrations of acid may lead to the formation of toxic chemicals that are mostly the result of lignin decomposition. Hydrolyzed lignocellulose products have been shown to include four primary types of inhibitors: acetic acid from the hemicellulose fraction, degradation of lignin products, sugar products, and extracts to be solubilized in pre-treatment.

On the other hand, fermentation inhibitors may be classified according to their origin:

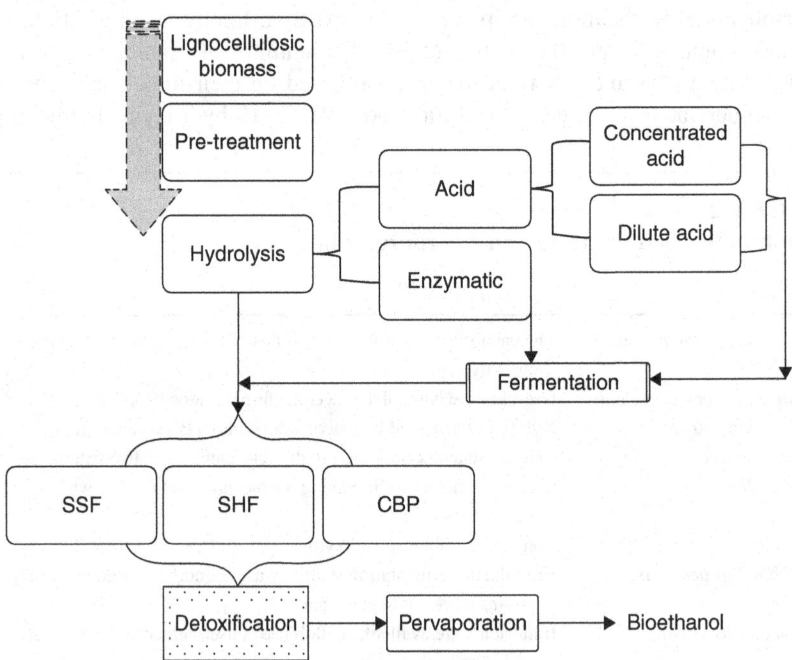

FIGURE 4.2 Schematic representation of bioethanol production from lignocellulosic biomass. SSF: solid state fermentation; SHF: Separate hydrolysis and fermentation and CBP: Consolidated biomass processing.

- extracts and acetic acid, such as aromatic, alcohols, and terpenes substances (i.e., tannins), are released during hydrolysis and pre-hydrolysis.
- inhibitors generated as a result of sugar degradation during pre-hydrolysis and hydrolysis (furfural, 5-HMF).
- lignin degradation products, which include large groups of polyaromatic and aromatic compounds with a wide range of components (syringaldehyde, phydroxybenzaldehyde, and cinnamaldehyde).
- acetic acid, ethanol, lactic acid, and glycerol are examples of fermentation products.
- Iron, nickel, chrome, and copper are examples of metals released by additives and equipment.
- lignin and acetic acid degradation products are the chemicals with the most inhibitory potential (Larsson et al., 1999).

To increase the fermentability, a detoxifying method would be required. They may be physical, chemical, or biological approaches, but they cannot be compared directly since the level of neutralization of inhibitors differs. The inhibitors are tolerated differently by various bacteria that are useful for this purpose. As a result, the most appropriate process is determined by the raw content used and the nature of the hydrolyzed products. Figure 4.2 illustrates the flow diagram for generating ethanol using lignocellulose as a raw material.

4.3 PROTOCOL FOR BIOETHANOL PRODUCTION

4.3.1 Enzymatic Hydrolysis of Starchy Feedstock

Enzymatic hydrolysis of cassava starch has been studied extensively. The amylase enzyme is used for conversion of starch into monosaccharide residues. An example of cassava starch has been taken as reference substrate for enzymatic saccharification (Ruiz et al., 2011). Following are the steps for starch hydrolysis:

- A 30% w/v Cassava starch was mixed with 0.8 g L^{-1} of Di-Ammonium Phosphate (DAP), 0.1 g L^{-1} Urea and 0.5 g L^{-1} $MgSO_4.7H_2O$.
- The pH of the reaction mixture was maintained at 5 using acetate buffer.
- The final reaction mixture was heated at 60°C for solubilization of starch.
- The alpha-amylase enzyme (Liquozyme® from Novozymes) was then added to reaction mixture. Enzyme was added at a concentration of 0.9 mg g-1 of alpha-amylase/starch.
- The reaction mixture was then incubated at 80°C for 180 min.
- The reducing sugars (dextrin) liberated from starch was estimated by DNS (3,5-di-nitro-salicylic acid) method.

$$\%\,\text{DE (Dextrin Equivalent)} = \frac{\text{g of reducing sugar expressed as glucose}}{\text{g of dry solid weight}} \times 100. \tag{4.7}$$

4.3.2 Enzymatic Hydrolysis of Lignocellulosic Biomass (LCB)

The LCB contains complex biopolymers such as cellulose, hemicellulose and lignin. Enzymatic hydrolysis of such biomass would require the action of cellulase, hemicellulase and β-glucosidase.

- A case study using the mulberry biomass for bioethanol production used treatment with vareious enzymes (Baksi et al., 2019). Similar approaches can be made for treatment of any LCB.
- The enzyme cocktail containing cellulase (1.3U mg^{-1}), hemicellulose (1.5U mg^{-1}) and β-glucosidase (7.7 U mg^{-1}) was used to treat LCB with a ratio of 1:1:2, respectively.
- All the enzymes were prepared in 30 mM citrate buffer of pH 4.5. Sodium azide of 0.1% w/v was added to the buffer to prevent microbial growth.
- The biomass and enzyme concentrations in the reaction mixture needs to be optimized.
- The biomass concentration and enzyme concentration of 125 g dry biomass L^{-1} and 17.8 g L^{-1}, respectively showed improved saccharification in case of mulberry biomass (Baksi et al., 2019).

4.3.3 Production of Bioethanol

Instruments required: Bioreactor, spectrophotometer and incubator shaker
Microorganism: *Saccharomyces cerevisiae*

Production media composition:

Yeast Extract	1% w/v
Peptone	2% w/v
Dextrose	10% w/v
Water	500 ml
pH	4.5

- Pre-inoculum was prepared by inoculating colonies of fresh grown culture of *Saccharomyces cerevisiae* into 50 ml working volume media in a 100ml conical flask.
- Cell growing at late log phase (OD at 660nm = 0.8) was inoculated in bioreactor with 500 ml working volume.
- Fermentation was carried out for 68 h under anaerobic conditions and the broth was further examined for ethanol production.

4.3.4 Estimation of Bioethanol

4.3.4.1 Colorimetric Method of Ethanol Estimation (Qi et al., 2011)

- The double enzyme-coupled assay was used for determination of ethanol concentration.

- Reagent I: Add Phenol (6 mmol L^{-1}) and EDTA (0.25 mmol L^{-1}, of pH7.5) in 0.1 M phosphate buffer of pH 7.5. Mixed it gently at room temperature.
- Reagent II: Enzyme cocktail of alcohol oxidase (3000U L^{-1}), peroxidase (600U L^{-1}) with aminoantipyrine (3·5 mmol L^{-1}) in 0.1 M phosphate buffer of pH 7.5.
- 100 μl of Reagent I was taken in a 96-well microtiter plate. Subsequently 10 μl ethanolic broth was added to it.
- In the above mixture, 100 μl of Reagent II was added and thoroughly mixed.
- The reaction mixture was then incubated in at 30°C for 60 min.
- The absorbance of the final product was monitored at 500 nm.
- A standard curve with ethanol was prepared using above mentioned double enzyme-coupled assay.
- The ethanol concentration in the broth was calculated from the standard curve.

4.3.4.2 Ethanol Estimation Using Gas Chromatography—Flame Ionization Detector (FID)

- Ethanol content in the fermentation broth was estimated via flame ionization detector (FID) (Roy et al., 2014).
- The injection port, the detector and programmed column had a temperature profile of 220°C, 240°C, and 130–175°C, respectively.
- A hydrogen and air mixture were used at a flow rate of 30 mL min^{-1} to ignite flame.
- A standard curve with ethanol was prepared using above mentioned program. The standard curve plotted with peak area vs concentration.
- The ethanol concentration in the broth was calculated from the standard curve.

CONCLUSIONS

Even though bioethanol is a viable substitute for fossil fuels with a minimal environmental effect, it is causing issues with the usage of raw materials like cereals, which are essential for the food sector. For biofuel production, expansion of agricultural area to cultivate grass for energy purposes implies competing with the food production crops. Several reports have been published to determine the demand for agricultural land to grow crops for ethanol production. The bioethanol production per hectare is mainly dependent on the crops utilized, but the average productivity in Europe (weighted by crop type) is now projected to be approximately 2790 liters/hectare (based on a 7 ton/hectare seed yield and 400 liters/ton). Alternatives are being researched, with a focus on revolutionary raw ingredients including *Miscanthus giganteus*, it is an inedible plant having high calorific value (roughly 4200 Kcal/kg dried matter), and filamentous fungi, i.e., *Trichoderma reesei*, which may also disintegrate complex lignocellulose molecule bonds.

REFERENCES

Baksi, Sibashish, Akash K. Ball, Ujjaini Sarkar, Debopam Banerjee, Alexander Wentzel, Heinz A. Preisig, Jagdish Chandra Kuniyal et al. "Efficacy of a novel sequential enzymatic hydrolysis of lignocellulosic biomass and inhibition characteristics of monosugars." *International Journal of Biological Macromolecules* 129 (2019): 634–644.

Balat, Mustafa, Havva Balat, and Cahide Öz. "Progress in bioethanol processing." *Progress in Energy and Combustion Science* 34, no. 5 (2008): 551–573.

Chandel, Anuj Kumar, E. S. Chan, Ravinder Rudravaram, M. Lakshmi Narasu, L. Venkateswar Rao, and Pogaku Ravindra. "Economics and environmental impact of bioethanol production technologies: an appraisal." *Biotechnology and Molecular Biology Reviews* 2, no. 1 (2007): 14–32.

Chum, Helena L., David K. Johnson, Stuart K. Black, and Ralph P. Overend. "Pretreatment-catalyst effects and the combined severity parameter." *Applied Biochemistry and Biotechnology* 24, no. 1 (1990): 1–14.

Da Silva, Rosangela, Renato Cataluna, Eliana Weber de Menezes, Dimitrios Samios, and Clarisse M. Sartori Piatnicki. "Effect of additives on the antiknock properties and Reid vapor pressure of gasoline." *Fuel* 84, no. 7–8 (2005): 951–959.

Di Nicola, Giovanni, Eleonora Santecchia, Giulio Santori, and Fabio Polonara. *Advances in the Development of Bioethanol: a Review*. IntechOpen, 2011.

Gírio, Francisco M., César Fonseca, Florbela Carvalheiro, Luís Chorão Duarte, Susana Marques, and Rafal Bogel-Łukasik. "Hemicelluloses for fuel ethanol: a review." *Bioresource Technology* 101, no. 13 (2010): 4775–4800.

Hahn-Hägerdal, Bärbel, Mats Galbe, Marie-F. Gorwa-Grauslund, Gunnar Lidén, and Guido Zacchi. "Bio-ethanol–the fuel of tomorrow from the residues of today." *Trends in Biotechnology* 24, no. 12 (2006): 549–556.

Lapuerta, Magin, Octavio Armas, and Reyes García-Contreras. "Stability of diesel–bioethanol blends for use in diesel engines." *Fuel* 86, no. 10–11 (2007): 1351–1357.

Larsson, Simona, Eva Palmqvist, Bärbel Hahn-Hägerdal, Charlotte Tengborg, Kerstin Stenberg, Guido Zacchi, and Nils-Olof Nilvebrant. "The generation of fermentation inhibitors during dilute acid hydrolysis of softwood." *Enzyme and Microbial Technology* 24, no. 3–4 (1999): 151–159.

Laser, Mark, Deborah Schulman, Stephen G. Allen, Joseph Lichwa, Michael J. Antal Jr, and Lee R. Lynd. "A comparison of liquid hot water and steam pre-treatments of sugar cane bagasse for bioconversion to ethanol." *Bioresource Technology* 81, no. 1 (2002): 33–44.

Lynd, Lee R., Willem H. Van Zyl, John E. McBride, and Mark Laser. "Consolidated bioprocessing of cellulosic biomass: an update." *Current Opinion in Biotechnology* 16, no. 5 (2005): 577–583.

Martinez, Diego, Randy M. Berka, Bernard Henrissat, Markku Saloheimo, Mikko Arvas, Scott E. Baker, Jarod Chapman et al. "Genome sequencing and analysis of the biomass-degrading fungus *Trichoderma reesei* (syn. *Hypocrea jecorina*)." *Nature Biotechnology* 26, no. 5 (2008): 553–560.

Ortiz, Rodomiro, Kenneth D. Sayre, Bram Govaerts, Raj Gupta, G. V. Subbarao, Tomohiro Ban, David Hodson, John M. Dixon, J. Iván Ortiz-Monasterio, and Matthew Reynolds. "Climate change: can wheat beat the heat?" *Agriculture, Ecosystems & Environment* 126, no. 1–2 (2008): 46–58.

Overend, Rolph P., and Esteban Chornet. "Fractionation of lignocellulosics by steam-aqueous pre-treatments." *Philosophical Transactions of the Royal Society of London. Series A, Mathematical and Physical Sciences* 321, no. 1561 (1987): 523–536.

Palmieri, Gianna, Paola Giardina, Carmen Bianco, Andrea Scaloni, Antonio Capasso, and Giovanni Sannia. "A novel white laccase from *Pleurotus ostreatus*." *Journal of Biological Chemistry* 272, no. 50 (1997): 31301–31307.

Palmqvist, Eva, and Bärbel Hahn-Hägerdal. "Fermentation of lignocellulosic hydrolysates. II: inhibitors and mechanisms of inhibition." *Bioresource Technology* 74, no. 1 (2000): 25–33.

Palmqvist, Eva, Bärbel Hahn-Hägerdal, Mats Galbe, and Guido Zacchi. "The effect of water-soluble inhibitors from steam-pretreated willow on enzymatic hydrolysis and ethanol fermentation." *Enzyme and Microbial Technology* 19, no. 6 (1996): 470–476.

Qi, X., Y. Zhang, R. Tu, Y. Lin, X. Li, and Q. Wang. "High-throughput screening and characterization of xylose-utilizing, ethanol-tolerant thermophilic bacteria for bioethanol production." *Journal of Applied Microbiology* 110, no. 6 (2011): 1584–1591.

Roy, Shantonu, M. Vishnuvardhan, and Debabrata Das. "Improvement of hydrogen production by newly isolated Thermoanaerobacterium thermosaccharolyticum IIT BT-ST1." *International Journal of Hydrogen Energy* 39, no. 14 (2014): 7541–7552.

Ruiz, Monica I., Clara I. Sánchez, Rodrigo G. Torrres, and Daniel R. Molina. "Enzymatic hydrolysis of cassava starch for production of bioethanol with a Colombian wild yeast strain." *Journal of the Brazilian Chemical Society* 22 (2011): 2337–2343.

Sánchez, Oscar J., and Carlos A. Cardona. "Trends in biotechnological production of fuel ethanol from different feedstocks." *Bioresource Technology* 99, no. 13 (2008): 5270–5295.

Shevchenko, S. M., K. Chang, J. Robinson, and J. N. Saddler. "Optimization of monosaccharide recovery by post-hydrolysis of the water-soluble hemicellulose component after steam explosion of softwood chips." *Bioresource Technology* 72, no. 3 (2000): 207–211.

Silverstein, Rebecca A., Ye Chen, Ratna R. Sharma-Shivappa, Michael D. Boyette, and Jason Osborne. "A comparison of chemical pre-treatment methods for improving saccharification of cotton stalks." *Bioresource Technology* 98, no. 16 (2007): 3000–3011.

Sun, Ye, and Jiayang Cheng. "Hydrolysis of lignocellulosic materials for ethanol production: a review." *Bioresource Technology* 83, no. 1 (2002): 1–11.

Wyman, Charles E. "Ethanol from lignocellulosic biomass: technology, economics, and opportunities." *Bioresource Technology* 50, no. 1 (1994): 3–15.

5 Biobutanol Production Process

5.1 INTRODUCTION

Production of liquid organic chemicals and fuels through the biorefinery concept has gained a lot of impetus. Butanol and ethanol are two popular liquid fuels that have been widely explored in recent times. Butanol is used as a solvent in many chemical industries and can also be used as an alternative fuel. In the early 19th century, butanol was produced from the fermentation of starchy substrate, mainly corn starch. Obligate anaerobes such as *Clostridium acetobutylicum* were exploited for butanol production. The considerable expansion of petrochemical infrastructure during this time led to the availability of butanol in larger quantities at a cheaper rate. Therefore, butanol production from the biological route became economically nonviable. Global warming, atmospheric pollution, and rises in crude oil prices have reignited the interest in the microbiological production of butanol.

The use of lignocellulosic biomass as feedstock for biobutanol production has made the entire process renewable in nature. Recently, technological advancement in the continuous fermentation of biomass and its integration with the alcohol extraction stage, also known as extractive fermentation, has helped considerably reduce the cost of operation. The acetone–butanol–ethanol (ABE) fermentation involves two phases. The acidogenic phase produces butyric, propionic, lactic, and acetic acids along with hydrogen and reducing equivalents. In the second phase, the hydrogen and reducing equivalents act as electron donor and solvent synthesis viz. butanol, acetone, ethanol, and isopropanol occur. Such metabolic pathway is prominently found in *C. acetobutylicum*.

A typical ABE fermentation has substantially low product (primarily butanol) yields (1.5% to 2.5% v/v). The probable reason for lower butanol yields could be because butanol itself is toxic to the microorganism producing it. The butanol concentration in 1–2% v/v inhibits microbial growth, leading to suboptimal yields. Butanol production by *C. acetobutylicum* showed yield and productivity of 1.5% and 4.5 g L^{-1} h^{-1}, respectively (Tigunova et al., 2013).

DOI: 10.1201/9781003224587-5

5.2 CHARACTERISTICS OF BUTANOL

Butanol, also known as *n*-butanol, is a colorless fluid having a pungent odor. It is a linear four aliphatic carbon alcohol with the molecular formula C_4H_9OH, and it also has three isomers, viz. isobutanol, fluorobutanol, and tertiary butanol. Butanol is soluble in organic solvents and partially soluble in water. It forms an azeotropic solution with water. Butanol is a comparatively low toxic compound with LD_{50} (Lethal Dose 50) value ranging between 2300–4400 mg/kg. However, among the primary alcohols, n-butanol has the highest toxicity (Lee et al., 2008). The main characteristics of butanol have been summarized in Table 5.1.

TABLE 5.1
The Physical and Chemical Characteristics of Butanol

Properties	Values
Melting point (°C)	–89.3
Molar mass, g/mol	74.12
Ignition temperature (°C)	35–37
Autoignition temperature (°C)	343–345
Flash point (°C)	25–29
Relative density (water: 1.0)	0.81
Critical pressure (hPa)	48.4
Critical temperature (°C)	287
Explosive limits (vol % in air)	1.4–11.3
Water solubility	9.0 mL per 100 mL
Relative vapor density (air: 1.0)	2.6
Vapor pressure (kPa at 20°C)	0.58
Boiling point (°C)	117–118
Density at 20°C (g/mL)	0.8098
Solubility in 100 g of water	–
Energy density (MJ L–1)	27–29.2
Energy/content value (BTU/gal)	110 000
Air–fuel ratio	11.2
Heat of vaporization (MJ/kg)	0.43
Liquid heat capacities (Cp) at STP (kJ/k mol K)	178
Research octane number	96
Motor octane number	78
Octanol/water partition coefficient (as logPo/w)*	0.88
Dipole moment (polarity)	1.66
Viscosity (10–3 Pa)	2.593

* LogP is a measure of hydrophobicity (lipophilicity) and is similar to polarity.

Source: Gholizadeh, 2010

5.3 TECHNOLOGIES FOR BUTANOL PRODUCTION

5.3.1 Chemical Synthesis of Butanol

Butanol has many advantages compared to ethanol, such as having higher energy potential hydrophobicity, and being non-hygroscopic and comparably less corrosive. The chemical synthesis process uses propylene or acetaldehyde for butanol production. Chemical processes like Oxo, Reppe, and crotonaldehyde hydrogenation are a few industrially commercialized processes.

In the Oxo synthesis (hydroformylation), carbon monoxide and hydrogen react with propylene carbon-carbon double bond in the presence of Co, Rh, or Ru catalysts. Different ratios of isomeric butanol are formed depending upon the reaction conditions (pressure and temperature) and the catalyst type.

The Reppe process is cost intensive and requires a low temperature and pressure. In this process, propylene reacts with carbon monoxide and water. At the same time, the crotonaldehyde hydrogenation process is the most widely commercialized. In this process, acetaldehyde undergoes aldol condensation followed by dehydration and hydrogenation to produce butanol. The other chemical synthesis methods use petroleum intermediates as feedstock for butanol production. In contrast, the crotonaldehyde hydrogenation process provides an alternative route of using ethanol derived from the fermentation of lignocellulosic biomass. The ethanol is dehydrated to acetaldehyde, which gets converted to butanol. By 2020, butanol's worldwide market volume had exceeded 5 million metric tons (Lee et al., 2008).

5.3.2 Biological Route for Butanol Synthesis: ABE Fermentation

Butanol is produced through a biological route known as acetone–butanol–ethanol (ABE)-type fermentation. The end products of the classical ABE fermentation process are acetone, butanol, and ethanol with a 3:6:1 ratio, respectively. The ABE fermentation is a sequential process where acid production is followed by alcohol formation phases. Once a suitable amount of short chain fatty acids (acetate and butyrate) are formed, this subsequently leads to the production of alcohol (ethanol and butanol).

The glucose gets metabolized through Embden Meyerhoff Parnas (EMP) pathway. In this pathway, glucose is oxidized into pyruvate. Concomitantly 2 moles of adenosine triphosphate (ATP) and 2 moles of nicotinamide dinucleotide (NADH) are produced per mole of glucose. In comparison, the pentose sugars are converted into pyruvate through the pentose phosphate pathway. In this pathway, 5 moles of ATP and NADH are used per 3 moles of pentose sugar. The fate of pyruvate depends upon the availability of oxygen. In anaerobic conditions, pyruvate gets converted to lactate with the help of lactate dehydrogenase, eventually regenerating NAD^+.

In another pathway, the pyruvate ferredoxin oxidoreductase converts pyruvate into acetyl-CoA, releasing 1 mole of CO_2 (Mitchell, 2001). The acetyl-CoA act as an intermediate branching point leading to acetate, butyrate, and ethanol synthesis.

During ethanol synthesis, the acetyl-CoA is converted into acetoaldehyde using acetoaldehyde dehydrogenase. This step consumes 1 mole of NADH, thereby regenerating NAD^+ and CoA. The alcohol dehydrogenase then converts acetaldehyde to ethanol by consuming 1 mole of NADH.

The acetate production from acetyl-CoA is catalyzed by two enzymes, namely acetyltransferase (phosphotransacetylase) and acetate kinase. The acetyl-CoA gets converted into acetylphosphate with the help of phosphotransacetylase. The acetate kinase then cleaves the phosphate group attached to acetylphosphate and forms acetate and ATP. Two moles of acetyl-CoA condense to form acetoacetyl-CoA in the presence of the thiolase enzyme. The acetoacetyl-CoA is another point of branching where it is channelized to synthesize acetone and butyrate. The acetoacetyl-CoA: acetate/butyrate: CoA transferase converts acetoacetyl-CoA to acetoacetate. This acetoacetate then gets converted to acetone with the help of acetoacetate decarboxylase, followed by the loss of 1 carbon as CO_2. The 3-hydroxybutyryl-CoA dehydrogenase converts acetoacetyl-CoA to 3 -hydroxybutyryl-CoA, consuming 1 mole of NADH. The 3- hydroxybutyryl-CoA is then catalyzed by crotonase to form crotonyl. The crotonyl is converted to butyryl-CoA using butyryl-CoA dehydrogenase and NADH.

The butyryl-CoA is another branching point where it can get converted into either butanol or butyrate depending on the fermentation phase (solvent production or acid production phase). The butyryl-CoA gets converted into butyraldehyde. The butyraldehyde gets converted to butanol with the help of butanol dehydrogenase and NADPH.

During acid production, the butyryl-CoA gets converted to butyrylphosphate in the presence of phosphate butyltransferase and phosphotransbutyrylase. The butyryl kinase enzyme then converts butyrylphosphate into butyrate along with ATP. It should be noted that isopropyl alcohol can be formed during solvent formation as a substitute for acetone. The isopropanol dehydrogenase converts acetone into isopropyl alcohol consuming one mole of NADPH. Under the classical ABE fermentation, during the acid production phase, 1 mole of glucose gets converted into 0.8 moles of butyrate, 0.4 moles of acetate, 2 moles of CO_2, and 2.4 moles of H_2 (Rogers and Gottschalk, 1993). During the solvent phase, one mole of glucose gets converted into 0.3 moles of acetone, 0.65 moles of butanol, 1.4 moles of H_2, and 2.3 moles of CO_2 (Rogers and Gottschalk, 1993).

One can control the acid and solvent phase duration if the concentrations of H_2, CO_2, organic acids, and mineral supplements are manipulated during ABE fermentation. Pyruvate gets converted into acetyl-CoA in the presence of ferredoxin oxidoreductases which is an iron-sulfur protein. Thus, supplementation with iron could be a critical process parameter for improving the yield of ABE metabolites. Alterations in iron concentrations can significantly affect the synthesis process for end products (Tashiro et al., 2005).

Depending on the substrate selection and culturing conditions, the ABE fermentation can take 48 h to 140 h. The concentration of solvents was 12–20 g/L after completing the process of periodic fermentation (Tashiro et al., 2005).

TABLE 5.2
ABE-type Clostridial Species

Microorganisms	Substrate
Clostridium butyricum	Starch, glucose, fructose, lactose, and xylose
C. tyrobutyricum	Cellulose, cellobiose, glucose, sucrose, and glycerin
C. beijerinckii	
C. pasterianum	
C. barkeri	
C. acetobutylicum	
C. thermobutyricum	
C. thermopalmarium	
C. sachharolyticum	
Butyribacterium methylotrophicum	Hexose, lactate, pyruvate, and CO, CO_2/H_2, and methanol
Pseudobutyrivibrio ruminis	Glucose

The genera *Clostridium, Butyrvibrio, Butyribacterium, Sarcina, Eubacterium, Fusobacterium*, and *Megasphera* are some reported strains in ABE-type fermentation. Among them, the most studied ones are the genera *Clostridium, Butyrvibrio*, and *Butyribacterium* (Zigova and Šturdík, 2000) (Table 5.2). Microorganisms of the genus *Clostridium* (*C. acetobutylicum, C. beijerinckii, C. saccharobutylicum*, and *C. saccharoperbutylacetonicum*) are used for the industrial-scale production of butanol. Clostridial species are strict anaerobes, rod-shaped, spore-forming, and gram-positive. They are isolated from soil, municipal organic waste, starchy foods, roots of nitrogen-fixing legumes, dairy products, etc. (Formanek et al., 1997). The *C. acetobutylicum* is the most reported strain for acetone and butanol production. Some supplementary byproducts are also synthesized along with butanol, viz. ethanol, lactate, and isopropanol. Moreover, *C. butyricum, C. tyrobutyricum, C. populeti, C. beijerinckii*, and *C. thermobutyricum* are known for butyric acid and butanol production. In contrast, *C. beijerinckii* has some contrasting metabolic profiles. It produces solvents similar in concentration when compared with *C. acetobutylicum*, but instead of acetone it synthesizes isopropanol (Huang, 2002). The *C. tetanomorphum* can produce an equimolar amount of butanol and ethanol and no other solvents (Jones and Woods, 1986).

The butanol concentrations at 1.0–2.0% are toxic to cells. It causes an impediment to cell growth, which eventually leads to poor growth and production efficiency (Oliinichuk et al., 2009). Butanol tinkers with the phospholipidic components cell membrane and thus disrupts its fluidity. This leads to hindered movement of sugar analogs into a cell. Butanol in a concentration of 6–12 g/L, when supplemented to an actively growing culture, led to an impediment of growth by 50%. At the same time, it increases the concentrations of acetone and ethanol (up to 40 g/L) (Paredes et al., 2005). The butanol toxicity can be mitigated by screening more tolerant strains or using molecular biology tools to improve the existing strains. Mutagenic strain improvement of *C. beijerinckii* NCIMB 8052 leads to the generation of robust mutants capable of tolerating higher concentrations of solvents.

Using mutagenesis, a new strain *C. beijerinckii* BA 101 was developed from *C. beijerinckii* NCIMB 8052 (Lee et al., 2008). In another example, the introduction of mutation in *lyt–1* gene in *C. acetobutylicum* ATCC 824 improved butanol concentrations by 5% (Liu et al., 2015). The gene-silencing technique using siRNA was used to downregulate specific genes, which helped improve butanol tolerance. The glycerine dehydrogenase gene (*gldA*) was downregulated using siRNA in *C. beijerinckii* NCIMB 8052, resulting in a 20% improvement in butanol tolerance (Liyanage et al., 2000).

5.4 FEEDSTOCKS FOR BIOBUTANOL PRODUCTION

The critical factor for the ABE fermentation process is the substrate's cost. Starch (corn, potato, wheat, manioc, etc.) and sugar-containing (sugarcane, beet, and molasses) energy crops have been used for the ABE process. The possibility of exploiting Clostridial species that can metabolize complex carbohydrates apart from mono and disaccharides, xylose and cellobiose, encouraged the researchers to explore cheaper feedstocks as substrates (Ezeji et al., 2007). Lignocellulosic biomass can be considered cheap, sustainable, and reliable feedstock for ABE-type fermentation. When cornmeal was used as a substrate by *C. beijerinckii* BA101, the production cost was reduced by 14.7%, while the butanol yield increased by 19.0% (0.4–0.52 g of butanol per gram of glucose) (Lee et al., 2008). The *C. beijerinckii* BA101, showed lower solvents yields using spray-dried soy molasses (SDSM) as substrate. However, supplementation with glucose in the SDSM media showed 113% (10.7–22.8 g/L) improvement in ABE yields (Qureshi et al., 2001). Solventogenesis was favored when SDSM was used as substrate and periodically supplemented with glucose or saccharose. Feedstock and Clostridial species involved in biobutanol production are given in Table 5.3.

Xylan-containing corn fibers were hydrolyzed, and its subsequent ABE fermentation was integrated into a single pot system, showing significantly high solvent yields of 44% w/w and productivity of 0.5 g L^{-1} h^{-1} (Qureshi and Li, 2006). In another study, continuous culture of *C. beijerinckii* P260 using wheat straw hydrolysate showed a 200% increment in butanol yields (Qureshi et al., 2007).

5.5 PURIFICATION AND SEPARATION OF BUTANOL

After fermentation, the liquid broth contains a mixture of butanol and other products. Since butanol forms an azeotropic mixture with water, it needs to be separated for further use. An extractive fermentation approach where fermentation is integrated with continuous butanol extraction has improved the product recovery and overall yield. Such technology has helped minimize the inhibitory effect of butanol and

TABLE 5.3
Feedstock and Clostridial Species Involved in Biobutanol Production

Substrate	Carbon source	Clostridial species	References
Grape pomace	Fructose, glucose, and sucrose	*Clostridium saccharobutylicum*	Law, 2010
Apple pomace	Fructose, glucose, and sucrose	*C. beijerinckii*	Voget et al., 1985
Jerusalem artichokes	Polyfructans	*C. beijerinckii*	Marchal, et al., 1986
Whey	Lactose	*C. acetobutylicum* *C. beijerinckii*	Maddox and Murray, 1983 Schoutens et al., 1984
Low-grade potatoes	Glucose and starch	*C. beijerinckii*	Nimcevic et al., 1998
Cane Molasses	Glucose and sucrose	*C. saccharobutylicum*	Ni et al., 2012
Soy molasses	Glucose and sucrose	*C. beijerinckii*	Qureshi et al.,2001
Miscanthus	Glucose	*C. beijerinckii*	Zhang and Ezeji, 2014
Wood hydrolysate	Glucose, mannose, and lignocellulose	*C. acetobutylicum*	Maddox and Murray, 1983
Peat	Glucose, xylose, and lignocellulose	*C. beijerinckii*	Forsberg et al., 1986
Palm oil effluent	Lignocellulose, oil, glucose, and xylose	*C. aurantibutyricum*	Sombrutai et al., 1996
Domestic organic waste	Lignocellulose, glucose, and xylose	*C. acetobutylicum*	López-Contreras et al., 2000

reduce the requirement of input energy. Gas stripping, pervaporation, and counter-current liquid extraction are the three prominent technologies for separating and purifying desired end products from the fermented broth (Figure 5.1).

Gas striping has been the most straightforward technology for purifying butanol from the broth. This process separates the fermentation product mixture using a counter-current mass and heat transfer between the vapor and liquid phase. The gas-striping technology has been integrated for both continuous and batch fermentation processes. The separation of the solvents from the fermentative broth was based on the principle of difference in boiling points of the solvent components. When fermentation was integrated with gas striping, improvements were shown in substrate conversion efficiency. This helped in performing fermentation at higher sugar concentrations (Ezeji et al., 2004).

The separation of butanol with a counter-current instrument is based on the difference in the coefficients of dispersion. Butanol readily solubilizes in organic solvents (extractants), but is sparingly soluble in fermentative broth. Therefore, butanol can be selectively condensed in extractants. Popular extractants used to isolate and purify butanol are decanol and oleic alcohol. A shielding membrane is used over the cultivation medium (the process of perstruction) to minimize the

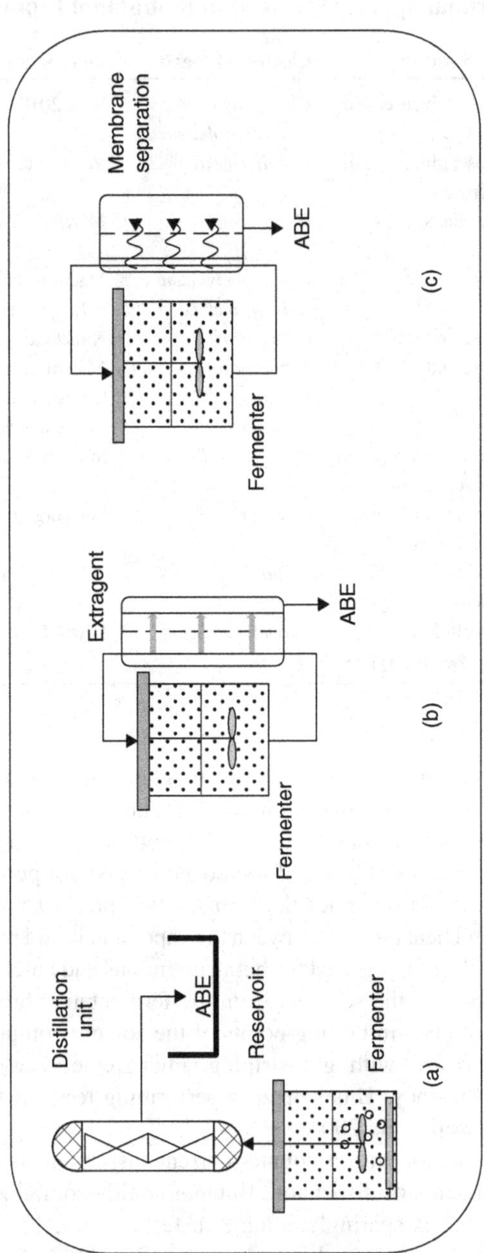

FIGURE 5.1 Schematic representation of technologies for production and recovery of butanol: (a) gas striping; (b) liquid extraction; and (c) pervaporation.

extractant's toxic impact on the cultures. High butanol accumulation rates were observed when a continuous culture of *C. acetobutylicum* was integrated with a liquid extraction process where decanol and oleic alcohol were used as extractants (Evans and Wang, 1988).

In the pervaporation process, a membrane separates components present in broth (evaporation through a membrane). In this process, fermented broth or the liquid mixture comes in contact with one side of the membrane, and the other side of the membrane is kept under a vacuum. The desired product passes through the membrane to the other side of the membrane (also known as permeate). On the other side, the permeate is subsequently removed in the form of vapors. Compared to other membrane separation technologies, pervaporation is a relatively cheap and less energy intensive process. Pervaporation is carried out using different methods to improve separation efficiencies, such as vacuum or carrier gas flow methods. The vacuum pervaporation method is prominently used in industry due to its simplicity in installation and minimal demand for high-end equipment.

There are three crucial and subsequent stages of pervaporation:

a. feed stream comes in contact with the membrane, selective sorption of the product occurs;
b. the product then diffuses through the membrane; and
c. on the vacuum side of the membrane, product desorption into the vapor phase takes place.

Various categories of membranes have been used for butanol purification. Polytetrafluoroethylene (PTFE) membrane polypropylene (PP) membrane, aqueous liquid membrane, and oleyl alcohol liquid membrane are some of the widely used membranes for butanol separation (Qureshi and Blaschek, 1999; Matsumura et al., 1998 and Liu and Feng, 2005). Membrane-based separation has helped improve the product recovery and yield by 50%.

5.6 PROTOCOL FOR BIOBUTANOL PRODUCTION

5.6.1 Cultivation Method for *Clostridium acetobutylicum* for Butanol Production

This section will discuss the cultivation method for a model organism, i.e., *Clostridium acetobutylicum* ATCC 824.

5.6.1.1 Growth Media Composition (Per Liter) (Baer et al., 1987)

60 g glucose, 0.5 g K_2HPO_4, 0.5 g KH_2PO_4, 2.2 g CH_3COONH_4, 0.2 g $MgSO_4 \cdot 7H_2O$, 0.01 g $MnSO_4 \cdot H_2O$, 0.01 g NaCl, 0.01 g $FeSO_4 \cdot 7H_2O$, 1 mg ρ-aminobenzoic acid, 1 mg thiamine, 0.01 mg biotin and 10 g cysteine HCl.

5.6.1.2 Process of Media Preparation and Fermentation

- Boil the double-distilled water for 10 min and cool it down to room temperature under N_2 gas bubbling.
- N_2 sparged water will be used for media preparation.
- Mix all the media components one by one in the prepared water.
- Immediately transfer the media into the fermenter under mild N_2 sparging.
- Proceed for in situ sterilization of the media.
- Add 5 to 20% v/v mid-log phase inoculum into the fermentation media.
- Fermentation was performed at 37°C for 120 h.

5.6.1.3 Product Recovery through Distillation (Roffler et al., 1988)

- Entire fermented broth, with *C. acetobutylicum*, was pumped onto the top of the extraction column having oleyl alcohol as extraction solvent (Figure 5.2).
- Using gravity flow, the extracted mixture was recycled back to the fermenter.
- The solvent gets collected at the top of the extraction column.
- The byproduct-laden extracting solvent was then channelized to a settling section to improve solvent broth disengagement.

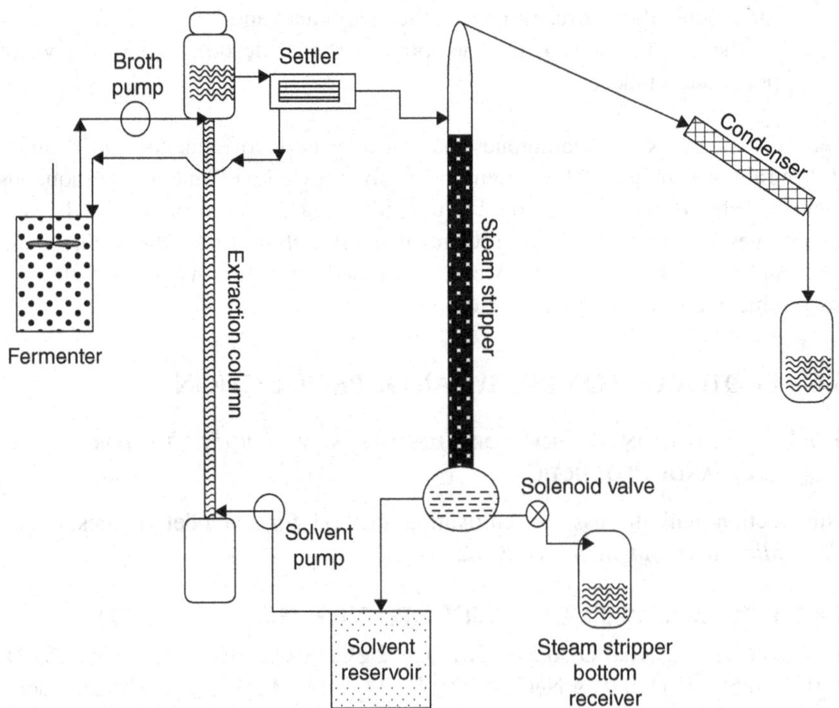

FIGURE 5.2 Schematic representation of steam stripper method for extraction of acetone and butanol.

- The solvent collected in the settling tank was then channelized to a steam stripper. The acetone, butanol, and ethanol were stripped from the extracting solvent (oleyl alcohol) and collected at the condenser.
- Broth getting collected in the settler tank was returned to the fermenter, while solvent was channelized to the steam stripper. Short chain alcohols such as butanol and ethanol were stripped from the extraction solvent and collected separately in a receiver.
- The extraction solvent can be recycled and used in successive operation cycles.

CONCLUSIONS

Butanol has various characteristics which make it an ideal candidate for its use as an alternative fuel. It has a higher calorific value when compared with ethanol and causes lesser damage to combustion engines. The bulk of butanol production is done through synthetic processes involving the usage of petroleum intermediates. The biological route of butanol synthesis has gained importance in recent years, as it can be produced from renewable sources (biomass-based feedstock). The costs of biomass (raw material) production, delivery, and storage has eventually influenced the overall cost of butanol production. Utilizing cellulosic and lignocellulosic biomass can certainly help minimize the operation cost and keep the process sustainable. Another challenge associated with biobutanol production is to separate the product from fermentation broth. The use of technologies viz. gas striping, liquid extraction, and pervaporation has undoubtedly improved the overall productivity and purity of the product.

REFERENCES

Baer, Shirley H., Hans P. Blaschek, and Terrance L. Smith. "Effect of butanol challenge and temperature on lipid composition and membrane fluidity of butanol-tolerant *Clostridium acetobutylicum*." *Applied and Environmental Microbiology* 53, no. 12 (1987): 2854–2861.

Evans, Patrick J., and Henry Y. Wang. "Enhancement of butanol formation by *Clostridium acetobutylicum* in the presence of decanol-oleyl alcohol mixed extractants." *Applied and Environmental Microbiology* 54, no. 7 (1988): 1662–1667.

Ezeji, T. C., Nasib Qureshi, and H. P. Blaschek. "Acetone butanol ethanol (ABE) production from concentrated substrate: reduction in substrate inhibition by fed-batch technique and product inhibition by gas stripping." *Applied Microbiology and Biotechnology* 63, no. 6 (2004): 653–658.

Ezeji, Thaddeus, Nasib Qureshi, and Hans P. Blaschek. "Butanol production from agricultural residues: impact of degradation products on *Clostridium beijerinckii* growth and butanol fermentation." *Biotechnology and Bioengineering* 97, no. 6 (2007): 1460–1469.

Formanek, Joseph, Roderick Mackie, and Hans P. Blaschek. "Enhanced butanol production by *Clostridium beijerinckii* BA101 grown in semidefined P2 medium containing

6 percent maltodextrin or glucose." *Applied and Environmental Microbiology* 63, no. 6 (1997): 2306–2310.

Forsberg, Cecil W., Herb E. Schellhorn, L. N. Gibbins, Frank Maine, and Eileen Mason. "The release of fermentable carbohydrate from peat by steam explosion and its use in the microbial production of solvents." *Biotechnology and Bioengineering* 28, no. 2 (1986): 176–184.

Gholizadeh, Laili. "Enhanced butanol production by free and immobilized *Clostridium* sp. cells using butyric acid as co-substrate." PhD diss, University of Borås, Sweden (2010).

Huang, Yu Liang, Zetang Wu, Likun Zhang, Chun Ming Cheung, and Shang-Tian Yang. "Production of carboxylic acids from hydrolyzed corn meal by immobilized cell fermentation in a fibrous-bed bioreactor." *Bioresource Technology* 82, no. 1 (2002): 51–59.

Jones, David T., and D. R. Woods. "Acetone-butanol fermentation revisited." *Microbiological Reviews* 50, no. 4 (1986): 484–524.

Law, Laurent. "Production of biobutanol from white grape pomace by *Clostridium saccharobutylicum* using submerged fermentation." PhD diss., Auckland University of Technology (2010).

Lee, Sang Yup, Jin Hwan Park, Seh Hee Jang, Lars K. Nielsen, Jaehyun Kim, and Kwang S. Jung. "Fermentative butanol production by Clostridia." *Biotechnology and Bioengineering* 101, no. 2 (2008): 209–228.

Liu, Fangfang, Li Liu, and Xianshe Feng. "Separation of acetone–butanol–ethanol (ABE) from dilute aqueous solutions by pervaporation." *Separation and Purification Technology* 42, no. 3 (2005): 273–282.

Liu, Zhen, Kai Qiao, Lei Tian, Quan Zhang, Zi-Yong Liu, and Fu-Li Li. "Spontaneous large-scale autolysis in *Clostridium acetobutylicum* contributes to generation of more spores." *Frontiers in Microbiology* 6 (2015): 950.

Liyanage, Hemachandra, Michael Young, and Eva R. Kashket. "Butanol tolerance of *Clostridium beijerinckii* NCIMB 8052 associated with down-regulation of gldA by antisense RNA." *Journal of Molecular Microbiology and Biotechnology* 2, no. 1 (2000): 87–93.

López-Contreras, Ana M., Pieternel A. M. Claassen, Hans Mooibroek, and Willem M. De Vos. "Utilisation of saccharides in extruded domestic organic waste by *Clostridium acetobutylicum* ATCC 824 for production of acetone, butanol and ethanol." *Applied Microbiology and Biotechnology* 54, no. 2 (2000): 162–167.

Maddox, I. S., and Anne E. Murray. "Production of n-butanol by fermentation of wood hydrolysate." *Biotechnology Letters* 5, no. 3 (1983): 175–178.

Marchal, R., M. Ropars, and J. P. Vandecasteele. "Conversion into acetone and butanol of lignocellulosic substrates pretreated by steam explosion." *Biotechnology Letters* 8, no. 5 (1986): 365–370.

Matsumura, M., H. Kataoka, M. Sueki, and K. Araki. "Energy saving effect of pervaporation using oleyl alcohol liquid membrane in butanol purification." *Bioprocess Engineering* 3, no. 2 (1988): 93–100.

Mitchell, Wilfrid J. "Biology and physiology." *Clostridia: Biotechnology and Medical Applications* (2001): 49–104.

Ni, Ye, Yun Wang, and Zhihao Sun. "Butanol production from cane molasses by *Clostridium saccharobutylicum* DSM 13864: batch and semicontinuous fermentation." *Applied Biochemistry and Biotechnology* 166, no. 8 (2012): 1896–1907.

Nimcevic, Dragan, M. Schuster, and James Richard Gapes. "Solvent production by *Clostridium beijerinckii* NRRL B592 growing on different potato media." *Applied Microbiology and Biotechnology* 50, no. 4 (1998): 426–428.

Oliinichuk, S., G. Kizyun, P. Shiyan, and V. Sosnits'kii. "Current and future technologies of biofuel production on the world market." *Kharchova Pererobna Prom.* 6, no. 358 (2009): 11–13.

Paredes, Carlos J., Keith V. Alsaker, and Eleftherios T. Papoutsakis. "A comparative genomic view of clostridial sporulation and physiology." *Nature Reviews Microbiology* 3, no. 12 (2005): 969–978.

Qureshi, N., A. Lolas, and H. P. Blaschek. "Soy molasses as fermentation substrate for production of butanol using *Clostridium beijerinckii* BA101." *Journal of industrial Microbiology and Biotechnology* 26, no. 5 (2001): 290–295.

Qureshi, Nasib, and Hans P. Blaschek. "Butanol recovery from model solution/fermentation broth by pervaporation: evaluation of membrane performance." *Biomass and Bioenergy* 17, no. 2 (1999): 175–184.

Qureshi, Nasib, Badal C. Saha, and Michael A. Cotta. "Butanol production from wheat straw hydrolysate using *Clostridium beijerinckii*." *Bioprocess and Biosystems Engineering* 30, no. 6 (2007): 419–427.

Qureshi, Nasib, Xin-Liang Li. "*Clostridium acetobutylicum*." *Biotechnology Progress* 22, no. 3 (2006): 673–680.

Roffler, S. R., H. W. Blanch, and C. R. Wilke. "In situ extractive fermentation of acetone and butanol." *Biotechnology and Bioengineering* 31, no. 2 (1988): 135–143.

Rogers, P., and G. Gottschalk. "Biochemistry and regulation of acid and solvent production in clostridia." *Biotechnology Series* (1993): 25–25.

Schoutens, G. H., M. C. H. Nieuwenhuizen, and N. W. F. Kossen. "Butanol from whey ultrafiltrate: batch experiments with *Clostridium beyerinckii* LMD 27.6." *Applied Microbiology and Biotechnology* 19, no. 4 (1984): 203–206.

Sombrutai, W., M. Takagi, and T. Yoshida. "Acetone-butanol fermentation by *Clostridium aurantibutyricum* ATCC 17777 from a model medium for palm oil mill eluent." *Journal of Fermentation and Bioengineering* 81 (1996): 543–547.

Tashiro, Yukihiro, Katsuhisa Takeda, Genta Kobayashi, and Kenji Sonomoto. "High production of acetone–butanol–ethanol with high cell density culture by cell-recycling and bleeding." *Journal of Biotechnology* 120, no. 2 (2005): 197–206.

Tigunova, O. A., S. M. Shulga, and Ya B. Blume. "Biobutanol as an alternative type of fuel." *Cytology and Genetics* 47, no. 6 (2013): 366–382.

Voget, C. E., C. F. Mignone, and R. J. Ertola. "Butanol production from apple pomace." *Biotechnology Letters* 7, no. 1 (1985): 43–46.

Zhang, Yan, and Thaddeus Chukwuemeka Ezeji. "Elucidating and alleviating impacts of lignocellulose-derived microbial inhibitors on *Clostridium beijerinckii* during fermentation of Miscanthus giganteus to butanol." *Journal of Industrial Microbiology and Biotechnology* 41, no. 10 (2014): 1505–1516.

Zigova, J., and E. Šturdík. "Advances in biotechnological production of butyric acid." *Journal of Industrial Microbiology and Biotechnology* 24, no. 3 (2000): 153–160.

6 Algal Cultivation and Biodiesel Production from Its Biomass

6.1 INTRODUCTION

Microalgae are microscopic, nonvascular, photoautotrophic organisms. They can produce oxygen by photosynthesis due to chlorophyll-a. They have been isolated from various habitats, such as fresh, brackish, or saline water bodies. They are generally found as solitary cells, showing almost no cellular differentiation. Interestingly, they have diverse metabolic pathways that can be exploited to synthesize valuable co-products, like potential biofuel, foods, feeds, and value-added bioactive molecules like amino acids, vitamins, carotenoids, and flavonoids. These value-added products have application in pharmaceutical, biorefinery-based industries, and agriculture. Microalgae offer several advantages over conventional terrestrial plants, viz., faster growth rate, can be grown in arid and barren lands, faster carbon dioxide sequestration, less energy requirement, and minimal nutritional requirement.

The algal biomass has been mooted as feedstock for the production of third-generation biofuel as it has 2–15-fold higher fatty acid content. Biodiesel, which is comparable to diesel, can be produced by exploiting microalgal biomass. The algal biomass accumulates high-energy content molecules, such as free fatty acids and triacylglycerols. These high-energy compounds can be chemically processed (using transesterification) into biodiesel. The fatty acids and TAGs profile vary from species to species and is greatly influenced by culture conditions. The carbon fixation ability of microalgae makes it a suitable candidate as a feedstock for biorefinery as it sequesters CO_2 emerging from industrial exhaust gases and can utilize nutrients from wastewaters to sustain their growth. Algal biomass has been used as feedstock to synthesize various types of bioenergy sources such as bioethanol, syngas, biomethane, and biohydrogen.

Despite being an excellent renewable energy source, commercial scale algal fuel production is at a naïve stage. This is due to higher operating costs, increased capital investment for algal farming, and lower productivity.

The problem of higher operating costs and lower productivity can be mitigated by the use of resources like industrial wastewater, sewage water, and farmland

DOI: 10.1201/9781003224587-6

runaway water. These resources can provide the nutritional requirements of algae and can eventually make algal cultivation sustainable.

At present, commercial algal farming uses four types of technology: lagoons, raceway or circular ponds, tubular photobioreactor, and heterotrophic cultivation. The photobioreactor technology gained more attention from a commercial viewpoint than others because of the following points:

- less prone to contamination;
- better control over growth parameters;
- large volumetric production due to its higher surface-to-volume ratio;
- minimum evaporation loss;
- higher yield compared to a raceway or open field cultivation.

Various types of photobioreactors have been designed and are operational at the pilot scale to produce value-added products. Still, their potential acceptance for commercial scale application is under evaluation.

There are certain disadvantages of tubular photobioreactors and has been mentioned below:

- overheating of the photobioreactor due to sunlight;
- homogenous mixing is difficult to achieve; thus, it is difficult to keep the temperature constant throughout the operation of the photobioreactor;
- the risk of photo-oxidation;
- extensive and sophisticated cleaning of equipment is complicated and thus increases the downtime.

The bottleneck, as mentioned above, needs to be addressed before scaling up the operation and its commercial realization.

6.2 PRESENT STATE OF THE ART OF PHOTOBIOREACTOR (PBR) DEVELOPMENT

Open system cultivation for microalgae is done mainly using ponds, open tanks, or raceways. It has a lower construction cost, needs less maintenance, and has lower energy input than closed bioreactors. Despite having crucial advantages, open system culture possesses a lot of disadvantages, like contamination by alien algae, grazing by zooplankton, less productivity, evaporative water loss, suboptimal culture condition, large area requirement, poor mixing, improper mixing, improper utilization of CO_2 and harvesting of light. In recent years, PBR development has evolved to use automated process control. Integrating engineering and biotechnology to overcome the drawbacks of photobioreactors has helped to realize its commercial operation. Research emphasis has been given towards developing vertical PBRs to abate photosaturation and photoinhibition. It also helped in improving photosynthetic efficiency and productivity. Implementation of the following strategies could improve algal biomass yields:

a) Providing adequate mixing in the reactor;
b) To provide an appropriate light-dark cycle to the cell to avert biofouling;
c) Improving high mass transfer efficiency for an adequate supply of CO_2 and forestall O_2;
d) Providing a larger surface area to volume ratio for improved volumetric output;
e) Precise control of temperature, pH, CO_2 flow, nutrient concentration.

The new PBR design with all the above features can work efficiently, but the major obstacle is less possibility of making it economically beneficial at a commercial scale. Some companies like ExxonMobilCrop, Chevron, Honeywell UOP, etc., are constructing PBR with innovative ideas at the experimental stage to scale up the biofuel production from algae.

Among the bioreactors, flat panel photobioreactors, tubular photobioreactors, raceway systems, and bubble column are operationalized at a moderate scale. New technology methods, such as the use of a sophisticated design map, and construction of a larger volume tank, made the PBR system suitable for algal farming. For example, treatment of municipal wastewater having 50,000 gallons/day capacity was designed using an offshore microalgal photobioreactor. It showed removal efficiency of 75%, 93% and 92% of total nitrogen, phosphorous and biological oxygen demand, respectively. Microalgal biomass yield of 3.5–22.7 g m^{-2} d^{-1} was also observed in this study (Novoveská et al., 2016). Most of the photobioreactors are designed with different specifications with respect to their working volume, material of construction, path length of light, and surface-to-volume ratio. Few standard features like the light energy conversion efficiently, loading ports for pouring culture media, CO_2 bubbling, mixing function for removal of accumulated oxygen, harvesting and improved mass transfer efficiency should be incorporated in designing PBR.

One of the most popular algal cultivation methods is the open raceway system (Figure 6.1). It is constructed with the basic principle of an oxidation pond with multiple paddles for better circulation and mixing (Figure 6.1a). The mixing

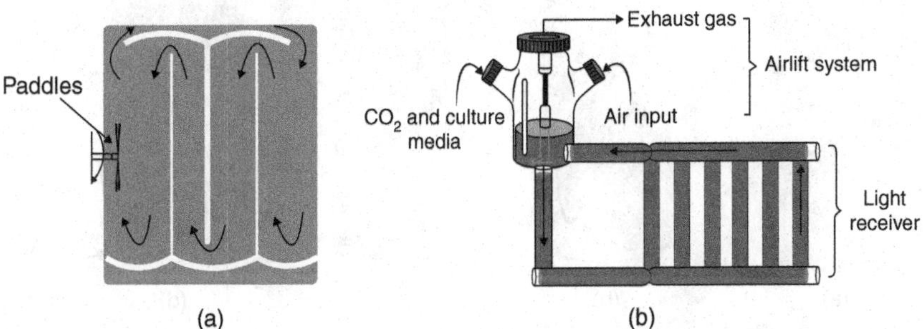

FIGURE 6.1 Schematic representation of a) raceway pond and b) horizontal column bioreactor.

done by paddles can provide improved energy efficiency as compared to aeration mediated mixing. Ironically, it suffers from lower yields because of prolonged exposure to light. Species with lower contamination risk are cultured in an open raceway system. Using a horizontal photobioreactor can minimize the risk of contamination (Figure 6.1b). A horizontal photobioreactor having a shorter light path length (5 cm) with an airlift pump with no contamination showed maximum biomass concentration and yield of 4.3 g L^{-1} and 0.11 g $m^{-2}d^{-1}$, respectively (Dogaris et al., 2015).

The vertical mixing can be achieved in column and flat plate photobioreactors (Figure 6.2), where the aeration promotes agitation and mixing. Homogenous mixing and efficient mass transfer can be achieved in vertical bubble column PBRs whereas the raceway system showed partial mixing (Figure 6.2a). Airlift column bioreactors are known for their efficient mixing (Figure 6.2b). Two interconnected zones within the airlift reactor are physically separated. The central column is considered a dark zone for upward flow, and the external side for downward flow is regarded as a light zone. The media circulation through the dark and light zone provides consistent light energy to all cells in the bioreactor.

For scaling up of vertical tubular configuration, the diameter of the column is increased, resulting in a decrement of surface-to-volume ratio (S/V ratio). This eventually leads to lower biomass yield. So harvesting and drying cells from such low biomass concentrations are cost-intensive processes. To avoid decrement in the S/V ratio, an annular configuration was introduced in the bubble column reactor (Figure 6.2c). The annular bioreactor's structure is actually a wrapped flat plate bioreactor that looks like a column bioreactor. The flat panel airlift (FPA) has a simple geometry that provides shorted pathlength for light, and S/V ratio also remains high (Figure 6.2d). As per a report, the FPA bioreactor showed 1.7 times greater production capacity than a conventional flat plate reactor using *Chlorella vulgaris* (Degen et al., 2001). The performance of a bioreactor is measured by its

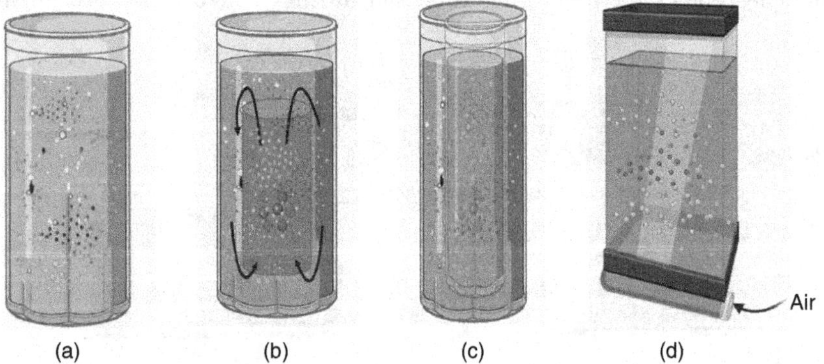

FIGURE 6.2 Schematic representation of a) bubble column, b) airlift, c) annular and d) flat plate photobioreactor.

net energy consumption. The energy balance between total energy produced by microalgae biomass (energy output) and energy required in biomass production is expressed as a net energy ratio (NER). The raceway system, in general, has a high NER ratio (>1.0) and good energy efficiency, whereas vertical reactors showed lower efficiency. Contrastingly, the horizontal PBR showed higher NER due to excessive energy consumption requirements for aeration.

A tubular bioreactor comprises a tube and pump system for the circulation of culture media. It is fitted with a degasser that eliminates the build-up of oxygen formed during photosynthesis. The advantage of such configuration is its high setting flexibility; it can be positioned horizontally, vertically, or in any other shape that is optimal for receiving light sources. However, during photolysis of water, oxygen is produced but is only partially removed from the system by the airlift reactor. Thus, oxygen shows its inhibitory effect in airlift reactor. Apart from the oxygen toxicity, the tubular configuration uses a lot of energy (>2500 W m^{-3} and NER = 0.2) to generate turbulence for proper gas/liquid mixing and mass transfer. In contrast, the raceway and flat plate systems use 3.72 W m^{-3} (NER = 8.34) and 53 W m^{-3} (NER = 4.51) for mixing and/or aeration, respectively. Operational challenges and advantages of different types of PBRs are mentioned in Table 6.1.

6.2.1 Photobioreactor Performance

Development and scale-up of microalgae cultivation are essential for the assessment of its economics and ecological viability. Cell density, biomass productivity, and product recovery are crucial yardsticks for accessing the performance of the PBR. Operational cost, energy expenditure, and construction capital investment determine the commercial viability of the process. Table 6.2 summarizes the different PBRs performance in terms of biomass yield.

6.2.1.1 Strategies to Improve the Photobioreactor Efficiency

Microalgae are considered the source of third-generation fuel producing feedstock and origin of valuable biochemical to meet the global population's needs. Microalgae cultivation industries can boost socio-economic development as microalgae have a positive impact on mitigating global climate change and improving food security. This necessitates the development of technologies related to microalgae cultivation for the larger interest of society. Manipulation of physiological responses of algae in a bioreactor can maximize yield and productivity. The basic requirements for microalgae growth are an adequate light source, carbon dioxide, and suitable nutrients to get the maximum output of biomass and bioactive compounds. The production rate is associated with the metabolic properties and given process conditions. The main aim is to get the highest productivity by increasing the photosynthetic rate and minimizing photo-oxidation.

A) Microalgae species selection: Each algal species has a different product profile of valuable compounds. The photobioreactor's target species should

TABLE 6.1
Bioreactor Configurations Used for Algal Cultivation

Production system	Advantages	Disadvantages	References
Open cultivation	Less cost-intensive and easy to maintain; Uses natural light source.	Prone to external contamination; marred with photoinhibition; severe evaporation loss; affected by climatic conditions.	Zhang et al., 2017
Ponds	Menial cost of construction. Easy maintenance.	Large land area is required; affected by climatic conditions. Poor light penetration at the bottom of pond; prone to external contamination.	Arora et al., 2018
Open tanks	Less cost-intensive and easy to maintain; process parameters are partially controllable.	Affected by climatic conditions; poor light penetration and uneven circulation; severe evaporation loss; Ineffective water usage.	Chen et al., 2018
Raceways Ponds	Less cost-intensive and easy to maintain; uses natural light source. Wasteland is used. Circulating algal cells harvest light better.	High energy consumption. Uneven light distribution. Photoinhibition might occur.	Hidasi and Belay, 2018
Close cultivation - photobioreactor	Easy to maintain pure culture; automated and controllable parameters.	High construction and maintenance cost. Outdoor reactors require cooling system.	
Annular/column	Improved light utilization efficiency; minimum photoinhibition. Proper aeration and nutrient distribution. Easy maintenance and operationality.	Annular construction is costly. Scale-up is difficult.	López-Rosales et al., 2016
Tubular	Large surface area to light; proper mixing.	High construction cost. Cleaning is difficult. Formation microalgae biofilm. Require lager space compared to other closed system.	Qin et al., 2018
Flat panel	Compared to other photobioreactors it is easy to construct.	Requires heat exchanging mechanism to maintain temperature. Scale-up is difficult.	Feng et al., 2011

have a fast growth rate to obtain the desired result. *Chlamydomonas* sp., for example, is known to make carbohydrates. Brown seaweeds and diatoms are used to extract fucoxanthin. Another factor to consider while choosing a good strain is the concentration of the incumbent product. The synthesis and quality of metabolites are determined by physiological factors and the biochemical makeup of the biomass. Different species utilize different values of light intensity and spectra in order to improve their productivity. The *Nannochloropsis* sp., a green microalga, produces a higher biomass when cultivated in blue light. Whereas, *Spirulina platensis*, produces lower photosynthetic pigments during the day and balances them at night. The following table shows the production capacity of different species of microalgae.

B) Aeration and mixing: To achieve a high photosynthetic yield and biomass yields, adequate carbon dioxide delivery and nutrient mixing throughout the culture media is required. Aeration facilitates mixing of water column moving along the algal cells to all regions of reactor tank, i.e., lighted area to darker area. As a result, algal cells shuttled back and forth between the two zones through a light and dark cycle, promoting faster growth and high biomass yield. Regulated mixing and proper supply of carbon dioxide and timely removal of oxygen must be monitored during the operation of PBRs.

C) Carbon dioxide and nutrients: Proper CO_2 provided to the microalgae culture effectively increases the rate of photosynthesis, thereby biomass yield. The global increase of CO_2 due to emissions from industry in terms of flue gas can be utilized in the PBRs, thus can help in mitigating the environmental issues associated with it. The CO_2 uptake can be enhanced by employing other growth factors to achieve a high amount of biomass.

The major determinants of algal growth are carbon, nitrogen, and phosphorous. Suitable concentrations of these three essential nutrients in the medium can enhance microalgae growth. Microalgae can successfully extract nutrients from nutrient-rich wastewater such as those from domestic sewage, the tannery industry, and aquaculture sludge. The absorption of ammoniacal nitrogen and phosphorus from tannery wastewater was highest. Groundwater could also be a good source of nutrients for microalgae cultivation as it is rich in minerals. Nutrient uptake in different microalgae cultures is given in the Table 6.3.

D) Light and temperature: Light is crucial for promoting microalgal growth and affects biomass production if intensity exceeds the maximum. A light field model for the prediction of light attenuation in bioreactor has been developed, which can easily be modified to accommodate various microalgae species in different photobioreactor configurations. The diverse light spectrum influences the microalgal photosynthetic rate. The LEDs emission spectra are a combination between blue (390–450 nm) and red (630–690 nm) and effectively increase microalgal biomass yield (Vadiveloo et al., 2015). A specific light spectrum can also increase the concentration of desired product to be synthesized by algal cells.

TABLE 6.2
Performance of Different Types of PBR Configurations

Microalgae	Reactor type	Maximum cell density ($g L^{-1}$)	Light intensity ($\mu E m^{-2} s^{-1}$)	References
Haematococcus pluvialis	Column	1.4	400	López-Rosales et al., 2016
Tetraselmis suecica	Column	1.16	1000	Zittelli et al., 2006
Phaeodactylum tricornutum	Column	4	1150	Mirón et al., 2002
Chlorella zofingiensis	Column	2.05	842	Zhu et al., 2013
Chaetoceros muelleri	Flat plate	1.5	805	Zhang and Richmond, 2003
Nannochloropsis sp.	Flat plate	2.430	1295	Cheng-Wu et al., 2001
Chlorella zofingiensis	Flat plate	0.680	1150	Feng et al., 2011
Spirulina platensis	Raceway	0.346	904	Jiménez et al., 2003
Chlorella sp.	Raceway	0.3	508	Hase et al., 2000
Chlorella sp.	Raceway	43	1181	Doucha and Lívanský, 2006
Anabaena sp.	Raceway	0.23	1099	Moreno et al., 2003
Scenedesmus obliquous	Raceway	0.810	822	Miranda et al., 2012
Muriellopsis sp.	Tubular	1.133	1258	Del Capo et al., 2001
Dunaliella salina	Tubular	0.6	255	García-González et al., 2005
Haematococcus pluvialis	Tubular	7.0	1200	López-Rosales et al., 2016
Synechocystis aquatilis	Tubular	1.0	1164	Ugwu et al., 2005
Chlorella sorokiniana	Tubular	1.5	688	Ugwu et al., 2002
Phaeodactylum tricornutum	Tubular	2.38	129	Fernández et al., 2001
Spirulina platensis	Tubular	3.4	674	Carlozzi, 2003
Phaeodactylum tricoruntum	Tubular	3.03	1135	Hall et al., 2003

Source: Yusoff et al., 2019

However, increasing light intensity is not always beneficial for all microalgae. Too much light can cause photoinhibition also. In dense culture, increasing light intensity causes light absorption and scattering that eventually reduces biomass yield. The availability of light to algal cells can be determined by measuring microalgal spectral absorption that can be used to predict and control the light transfer and biomass production in a photobioreactor (Kandilian et al., 2016). Solar energy is the best alternative source of light. If it is exploited efficiently, it can decrease the costs of operation of PBRs. It is not a reliable light source because of its temporal changes based on seasonal variation. Aside from that, artificial light sources such as high intensity discharged light (HID) and light-emitting diode (LED), while less expensive, provide a steady source throughout the seasons. Biomass productivity of 0.211 g L^{-1} d^{-1} at a high light intensity of 182.5 μmol m^{-2} $s^{-1}$was observed when *Scenedesmus* was cultivated in tannery wastewater (da Fontoura et al., 2017). The strategies and performance of the light system used in the photoreactor system are shown in Table 6.4.

TABLE 6.3
Nutrition Uptake by Microalgae

Culture system	Microalgae species	Nutrients sources	Nutrient uptake rates, total nitrogen (TN), total phosphorus (TP)	Microalgae biomass/product yield	References
Flask batch culture	*Scenedesmus* sp.	Tannery wastewater	Total ammonia 86%; soluble reactive phosphorus 97%	0.9 g L^{-1}	da Fontoura et al., 2017
Simultaneous biogas production and nutrient reduction system	*Chlorella* sp.	Biogas slurry nutrients	TN 81%; TP 80%	0.5 g L^{-1}	Yan et al., 2016
Fed-batch cultivation	*Arthrospira platensis*	Substrates (sodium glutamate) as metabolic stress and nitrate feeding strategy	Nitrate reduction, >200%	Algae biomass—8.0 g L^{-} [1]Phycocyanin—0.34 mg mL^{-1}	Manirafasha et al., 2018
Column reactors	*Ettlia* sp.	Ground water high in nutrients, N and P	P removal rate—6.0 mg L^{-1} $_{d}^{-1}$ N removal rate—11.0 mg L^{-1} $_{d}^{-1}$	Algae biomass, 1.0–1.4 g L^{-1}	Rezvani et al., 2017
Tubular airlift bioreactors	*Nannochloropsis* sp.	Anaerobic digestion of food waste	na	Algae biomass, 0.3–0.4 g L^{-1}	Mayers et al., 2017

Source: Yusoff et al., 2019

6.3 PROTOCOL FOR ALGAL CULTIVATION

The *Chlorella* is a microscopic, nonvascular, spherical (2–10 μm in diameter), eukaryotic algae found in single or in chain form. It obtains nutrients via autotrophic, heterotrophic, and mixotrophic mode of nutrition. The cultivation method of this species includes the open pond system where raceway types of ponds or lakes are suitable, but it suffers from high contamination problems. The major advantage of this type of cultivation system is its cheaper construction cost and exploitation of unusual conditions. Another method is the use of photobioreactor, a closed

TABLE 6.4
The Strategies and Performance of the Light System in PBRs

Type of light source	Advantages	References
Use of light-limiting diodes (LED)	Improved biomass productivity and synthesis of value-added products (carotenoids and phycocyanin)	Lima et al., 2018
Flashing lighting or dynamic light condition using internal LED	Reactor can be scaled up to greater extent; Prevention from photoinhibition, Less thermal dissipation; Low xanthophyll production.	Hu and Sato,2017
Light and CO_2 synergy	Enhancement in biomass and lipid yields.	Seo et al., 2017
The green solar collector (GSC)—use lenses and light guides	Efficient capturing mechanism of solar energy, reduced operation cost.	Zijffers et al., 2008
Mechanically stirred bioreactor	Better mass transfer; Homogenous distribution of cells.	Zhang., 2013
Photobioreactor equipped with a double jacket and a lower lighting (LED) panel	Improved biomass yields.	Brzychczyk et al., 2020
Use of blue and red spectra	Increases the photosynthetic efficiency of the algal cells.	Schulze et al., 2016
Light in immobilized cell cultures	Microalgae cell immobilized in agar gel to reduce contamination and easy metabolite recovery.	Kandilian et al., 2017
Use of optical fibers	Light can be transferred to the interior parts of the bioreactors where incident light cannot reach.	Sun et al., 2016

system where light is provided externally with all the necessary nutrients to attain maximum growth. The maximum theoretical photosynthetic efficiency is one of the basic insurmountable constraints towards algal growth. It depends on incident photon number and the number of photons absorbed per unit surface area per unit time. The following formula can be used to measure photosynthetic photon flux density (1 μ mole = 6.023×10^{17} photons).

Materials required:

A) Instruments: photobioreactor, gas chromatography, Quanta meter, column, spectrophotometer.
B) Medium: define/synthetic media is used
C) Water is added for volume makeup to 2.5 L

TAP (tris-acetate-phosphate medium)	2.5 L
TAP salt stock solution	12.5 ml L^{-1}(v/v) [31.25ml in 2.5L]
Phosphate stock solution	0.375 ml L^{-1} (v/v) [0.94ml in 2.5L]
Hunter trace metals	1 ml L^{-1} (v/v) (2.5ml in 2.5L)
Vitamin stock solution	1 ml L^{-1} (v/v) (2.5ml in 2.5L)
Glacial acetic acid	1 ml L^{-1} (v/v) (2.5ml in 2.5L)
Tris base	2.42 g L^{-1} (6.05g in 2.5L)
Addition of buffer to maintain pH	

Methodology:

A) Culture media preparation:
- A single colony from a previously grown Petri dish is isolated as inoculum, where the microorganism should be active to get active culture after subculture.
- At the mid log phase, the culture is transferred in a fresh media.
- The culture is then allowed to grow in the airlift reactor at the following conditions (Figure 6.3).

  Temperature = 25°C

  Light intensity = 137.11 μ mole m^{-2} s^{-1}

  Initial pH of the medium is 7.2

  Being a slow-growing algae, the sample must be withdrawn at 12 hours of interval.

B) Sample collection method:
- After the closing of the exhaust filter, the sample port is opened to collect the sample.
- Sample moves out slowly as the high pressure is generated in the reactor and 2 ml of sample is collected from the port.

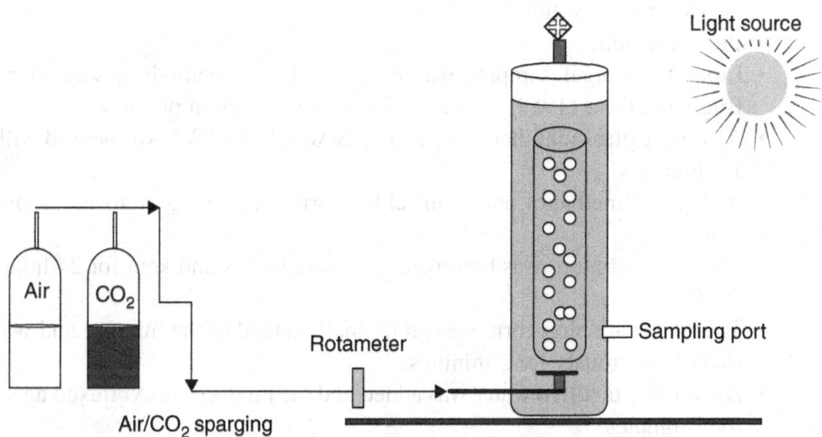

FIGURE 6.3 Schematic representation of airlift reactor setup for algal cultivation.

- Soon after taking the sample, the sample port is closed, and the exhaust air filter is to be opened.
- The same is repeated at 12-hour intervals; in each case, 2–3 ml of sample is drained out first to avoid the collection of previous samples accumulated at the interface of the port.
- Collected samples are centrifuged at 6000 rpm for 15 minutes, the supernatant is separated and the pellet collected for the next test.
- One set of samples is collected for determination of chlorophyll content and the remaining set is used to measure dry cell weight and analysis of substrate concentration.

C) Chlorophyll content determination:
- 1 ml of 95% methanol is mixed with the cell pellet and stored overnight.
- The suspension is centrifuged as the absorbance of the supernatant is taken at 650 nm, and 665 nm against 95% methanol is blank.
- Chlorophyll concentration estimation is done by a standard formula.

D) Dry cell weight determination:
- The weight of the vacant Eppendorf is taken each time before taking the sample.
- All samples are centrifuged at 6000 rpm for 15 minutes, and the supernatant is separated to get the cell mass as pellet.
- The pellets collected are washed with distilled water to remove traces of salts, minerals, and any media components present in the pellet; after that, the water is drained off.
- The collected pellet is again mixed with distilled water and centrifuged at 6000 rpm for 10 minutes the mixture is stored overnight to dry at 60^{0}C.
- The Eppendorf tube with the pellet is weighed, and the difference between the pellet containing Eppendorf and the empty Eppendorf is calculated to get the dry cell weight.

E) Lipid extraction:
- Lipid from algal samples was extracted using methods described by Bligh and Dyer (1959). It is a liquid–liquid extraction process.
- An empty glass vial that was previously weighed as W1 was poured with 1 g biomass.
- A 3. 2 ml methanol and 1 ml chloroform were added into the above vial.
- The entire mixture was homogenized using vortex and kept for 24 hours at 25°C.
- Then 1 ml of chloroform was additionally added to the mixture and was shaken vigorously for 5 minutes.
- A 1.8 ml of distilled water was added and the mixture was vortexed again for 2 minutes.
- The organic layer was separated by centrifuging the mixture for 10 min at 2000 rpm.

- The lower organic layer was filtered through Whatman No.1 filter paper into a clean bottle.
- The solvent was evaporated using hot water bath at 104°C for 1 h.
- The final weight of the vial was again recorded as W2.
- By subtracting the weight of W1 from W2, the lipid content can be estimated.

F) Fatty acid methyl esterification of lipid (Ichihara and Yumeto, 2010):

- Into the extracted lipid, 1 ml of 1 M KOH in 70% ethanol was added and refluxed at 90°C for 1 h.
- The entire mixture was cooled down to room temperature and 0.2 ml of 6 M HCl was added in the above mixture. This leads to acidification of the mixture.
- After acidification, 1 ml of water was added followed by 1 ml of hexane into the mixture.
- The final mixture was vortexed for 2 min and then allowed for phase separation.
- The organic layer containing the fatty acids was collected.
- The hexane was removed through evaporation.
- The fatty acids were methylated by adding 1 ml of 10% BF_3 in methanol at 37°C for 20 min.
- A 2ml water was added to methylated mixture followed by 2 ml of hexane to extract the fatty acid methyl esters (FAME).
- The FAME sample was analyzed by gas chromatography fitted with Omegawax-250 capillary column using a flame ionization detector (FID). Chromatographic peaks were identified by comparing to a FAME standard mixture.

G) Substrate consumption analysis by gas chromatography—flame ionization detection:

- The concentration of acetic acid present in the supernatant is measured using GC with an FID and capillary column coated with 10% PEG -20M and 2% H_3PO_4.
- The temperatures of the injection port, oven, detector and program are fixed to 220ºC, 240ºC, 240ºC and 130–175ºC respectively.
- Carrier gas (N_2) flow rate is maintained at 20mL min^{-1}. Hydrogen and air mixture flow rate is controlled at 30ml/min needed to generate a flame.

  The chromatogram is prepared by plotting detector response against retention time which provides a spectrum of the peak for an analyte present in the sample. The area under a peak is proportional to the amount of analyte, which indicates the amount of substrate present at a particular time.

CONCLUSIONS

Microalgae are autotrophic organism that needs minimal nutritional inputs for their growth. Cultivation of algae has been done in various types of photobioreactors.

Photobioreactors provide various optimized process parameters that are necessary for sustainable growth. Different configurations of PBRs provide vivid advantages and disadvantages. Open cultivation is the simplest and most cost-effective way of algal biomass production, but the chances of external contamination mar its long-time operation.

Moreover, raceway ponds are less cost-intensive, and continuous circulation of algal cells due to paddled agitation improves light-harvesting efficiency. To negate the problem of external contamination, the advent of closed photobioreactors came into prominence. Such bioreactors provided greater mass transfer efficiency, along with enhanced light penetration. Difficulty with *in-situ* cleaning and the forming of algal deposition on the bioreactor wall are disadvantages of closed photobioreactors. Algae can be grown using various wastewater viz. tannery waste, municipal wastes, etc. This eventually can create a win-win situation where waste treatment can be realized concomitantly with algal biomass production.

REFERENCES

Arora, Neha, Alok Patel, Parul A. Pruthi, Krishna Mohan Poluri, and Vikas Pruthi. "Utilization of stagnant non-potable pond water for cultivating oleaginous microalga *Chlorella minutissima* for biodiesel production." *Renewable Energy* 126 (2018): 30–37.

Bligh, E. Graham, and W. Justin Dyer. "A rapid method of total lipid extraction and purification." *Canadian Journal of Biochemistry and Physiology* 37, no. 8 (1959): 911–917.

Brzychczyk, Beata, Tomasz Hebda, and Norbert Pedryc. "The Influence of Artificial Lighting Systems on the Cultivation of Algae: The Example of Chlorella vulgaris." *Energies* 13, no. 22 (2020): 5994.

Carlozzi, Pietro. "Dilution of solar radiation through 'culture' lamination in photobioreactor rows facing south–north: a way to improve the efficiency of light utilization by cyanobacteria (*Arthrospira platensis*)." *Biotechnology and Bioengineering* 81, no. 3 (2003): 305–315.

Chen, Bang-Fuh, Hung-Kai Yang, Chih-Hua Wu, Tzu-Chiang Lee, and Bing Chen. "Numerical study of liquid mixing in microalgae-farming tanks with baffles." *Ocean Engineering* 161 (2018): 168–186.

Cheng-Wu, Zhang, Odi Zmora, Reuven Kopel, and Amos Richmond. "An industrial-size flat plate glass reactor for mass production of *Nannochloropsis* sp. (Eustigmatophyceae)." *Aquaculture* 195, no. 1–2 (2001): 35–49.

da Fontoura, Juliana Tolfo, Guilherme Sebastião Rolim, Marcelo Farenzena, and Mariliz Gutterres. "Influence of light intensity and tannery wastewater concentration on biomass production and nutrient removal by microalgae *Scenedesmus* sp." *Process Safety and Environmental Protection* 111 (2017): 355–362.

Degen, Jörg, Andrea Uebele, Axel Retze, Ulrike Schmid-Staiger, and Walter Trösch. "A novel airlift photobioreactor with baffles for improved light utilization through the flashing light effect." *Journal of Biotechnology* 92, no. 2 (2001): 89–94.

Del Campo, José A., Herminia Rodrıguez, José Moreno, M. Angeles Vargas, Joaquın Rivas, and Miguel G. Guerrero. "Lutein production by *Muriellopsis* sp. in an outdoor tubular photobioreactor." *Journal of Biotechnology* 85, no. 3 (2001): 289–295.

Dogaris, Ioannis, Michael Welch, Andreas Meiser, Lawrence Walmsley, and George Philippidis. "A novel horizontal photobioreactor for high-density cultivation of microalgae." *Bioresource Technology* 198 (2015): 316–324.

Doucha, Jirí, and Karel Lívanský. "Productivity, CO_2/O_2 exchange and hydraulics in outdoor open high density microalgal (*Chlorella* sp.) photobioreactors operated in a Middle and Southern European climate." *Journal of Applied Phycology* 18, no. 6 (2006): 811–826.

Feng, Pingzhong, Zhongyang Deng, Zhengyu Hu, and Lu Fan. "Lipid accumulation and growth of *Chlorella zofingiensis* in flat plate photobioreactors outdoors." *Bioresource Technology* 102, no. 22 (2011): 10577–10584.

Fernández, FG Acién, JM Fernández Sevilla, JA Sánchez Pérez, E. Molina Grima, and Y. Chisti. "Airlift-driven external-loop tubular photobioreactors for outdoor production of microalgae: assessment of design and performance." *Chemical Engineering Science* 56, no. 8 (2001): 2721–2732.

García-González, M., J. Moreno, J. P. Canavate, V. Anguis, A. Prieto, C. Manzano, F. J. Florencio, and M. G. Guerrero. "Conditions for open-air outdoor culture of *Dunaliella salina* in southern Spain." *Journal of Applied Phycology* 15, no. 2 (2003): 177–184.

Hall, David O., F. Guerrero Acién Fernández, E. Cañizares Guerrero, K. Krishna Rao, and E. Molina Grima. "Outdoor helical tubular photobioreactors for microalgal production: modeling of fluid-dynamics and mass transfer and assessment of biomass productivity." *Biotechnology and Bioengineering* 82, no. 1 (2003): 62–73.

Hase, Ryouetsu, Hiroyoshi Oikawa, Chiyo Sasao, Masahiko Morita, and Yoshitomo Watanabe. "Photosynthetic production of microalgal biomass in a raceway system under greenhouse conditions in Sendai city." *Journal of Bioscience and Bioengineering* 89, no. 2 (2000): 157–163.

Hidasi, Nora, and Amha Belay. "Diurnal variation of various culture and biochemical parameters of *Arthrospira platensis* in large-scale outdoor raceway ponds." *Algal Research* 29 (2018): 121–129.

Hu, Jin-Yang, and Toru Sato. "A photobioreactor for microalgae cultivation with internal illumination considering flashing light effect and optimized light-source arrangement." *Energy Conversion and Management* 133 (2017): 558–565.

Ichihara, Ken'ichi, and Yumeto Fukubayashi. "Preparation of fatty acid methyl esters for gas-liquid chromatography [S]." *Journal of Lipid Research* 51, no. 3 (2010): 635–640.

Jiménez, Carlos, Belén R. Cossío, Diego Labella, and F. Xavier Niell. "The feasibility of industrial production of Spirulina (Arthrospira) in Southern Spain." *Aquaculture* 217, no. 1–4 (2003): 179–190.

Kandilian, Razmig, Antoine Soulies, Jérémy Pruvost, Benoit Rousseau, Jack Legrand, and Laurent Pilon. "Simple method for measuring the spectral absorption cross-section of microalgae." *Chemical Engineering Science* 146 (2016): 357–368.

Kandilian, Razmig, Bruno Jesus, Jack Legrand, Laurent Pilon, and Jérémy Pruvost. "Light transfer in agar immobilized microalgae cell cultures." *Journal of Quantitative Spectroscopy and Radiative Transfer* 198 (2017): 81–92.

Lima, Gustavo M., Pedro CN Teixeira, Cláudia MLL Teixeira, Diego Filócomo, and Celso LS Lage. "Influence of spectral light quality on the pigment concentrations and biomass productivity of *Arthrospira platensis*." *Algal Research* 31 (2018): 157–166.

López-Rosales, L., F. García-Camacho, A. Sánchez-Mirón, E. Martín Beato, Yusuf Chisti, and E. Molina Grima. "Pilot-scale bubble column photobioreactor culture of a marine dinoflagellate microalga illuminated with light emission diodes." *Bioresource technology* 216 (2016): 845–855.

Manirafasha, Emmanuel, Theophile Murwanashyaka, Theoneste Ndikubwimana, Nur Rashid Ahmed, Jingyi Liu, Yinghua Lu, Xianhai Zeng, Xueping Ling, and Keju Jing. "Enhancement of cell growth and phycocyanin production in Arthrospira (Spirulina) platensis by metabolic stress and nitrate fed-batch." *Bioresource Technology* 255 (2018): 293–301.

Mayers, Joshua J., Anna Ekman Nilsson, Eva Albers, and Kevin J. Flynn. "Nutrients from anaerobic digestion effluents for cultivation of the microalga Nannochloropsis sp.—impact on growth, biochemical composition and the potential for cost and environmental impact savings." *Algal Research* 26 (2017): 275–286.

Miranda, J. R., Paula C. Passarinho, and Luisa Gouveia. "Bioethanol production from *Scenedesmus obliquus* sugars: the influence of photobioreactors and culture conditions on biomass production." *Applied Microbiology and Biotechnology* 96, no. 2 (2012): 555–564.

Mirón, Asterio Sánchez, Marie-Carmen Cerón García, Francisco García Camacho, Emilio Molina Grima, and Yusuf Chisti. "Growth and biochemical characterization of microalgal biomass produced in bubble column and airlift photobioreactors: studies in fed-batch culture." *Enzyme and Microbial Technology* 31, no. 7 (2002): 1015–1023.

Moreno, José, M. Ángeles Vargas, Herminia Rodrıguez, Joaquın Rivas, and Miguel G. Guerrero. "Outdoor cultivation of a nitrogen-fixing marine cyanobacterium, *Anabaena* sp. ATCC 33047." *Biomolecular Engineering* 20, no. 4–6 (2003): 191–197.

Novoveská, Lucie, Anastasia KM Zapata, Jeffrey B. Zabolotney, Matthew C. Atwood, and Eric R. Sundstrom. "Optimizing microalgae cultivation and wastewater treatment in large-scale offshore photobioreactors." *Algal Research* 18 (2016): 86–94.

Qin, Chao, Yuling Lei, and Jing Wu. "Light/dark cycle enhancement and energy consumption of tubular microalgal photobioreactors with discrete double inclined ribs." *Bioresources and Bioprocessing* 5, no. 1 (2018): 1–10.

Rezvani, Fariba, Mohammad-Hossein Sarrafzadeh, Seong-Hyun Seo, and Hee-Mock Oh. "Phosphorus optimization for simultaneous nitrate-contaminated groundwater treatment and algae biomass production using Ettlia sp." *Bioresource Technology* 244 (2017): 785–792.

Schulze, Peter SC, Hugo GC Pereira, Tamára FC Santos, Lisa Schueler, Rui Guerra, Luísa A. Barreira, José A. Perales, and João CS Varela. "Effect of light quality supplied by light emitting diodes (LEDs) on growth and biochemical profiles of *Nannochloropsis oculata* and *Tetraselmis chuii*." *Algal Research* 16 (2016): 387–398.

Seo, Seong-Hyun, Ji-San Ha, Chan Yoo, Ankita Srivastava, Chi-Yong Ahn, Dae-Hyun Cho, Hyun-Joon La, Myung-Soo Han, and Hee-Mock Oh. "Light intensity as major factor to maximize biomass and lipid productivity of *Ettlia* sp. in CO_2-controlled photoautotrophic chemostat." *Bioresource Technology* 244 (2017): 621–628.

Sun, Yahui, Yun Huang, Qiang Liao, Qian Fu, and Xun Zhu. "Enhancement of microalgae production by embedding hollow light guides to a flat-plate photobioreactor." *Bioresource Technology* 207 (2016): 31–38.

Ugwu, C. U., J. C. Ogbonna, and H. Tanaka. "Light/dark cyclic movement of algal culture (Synechocystis aquatilis) in outdoor inclined tubular photobioreactor equipped with static mixers for efficient production of biomass." *Biotechnology Letters* 27, no. 2 (2005): 75–78.

Ugwu, C., J. Ogbonna, and H. Tanaka. "Improvement of mass transfer characteristics and productivities of inclined tubular photobioreactors by installation of internal static mixers." *Applied Microbiology and Biotechnology* 58, no. 5 (2002): 600–607.

Vadiveloo, Ashiwin, Navid R. Moheimani, Jeffrey J. Cosgrove, Parisa A. Bahri, and David Parlevliet. "Effect of different light spectra on the growth and productivity of acclimated *Nannochloropsis sp.* (Eustigmatophyceae)." *Algal Research* 8 (2015): 121–127.

Yan, Cheng, Raúl Muñoz, Liandong Zhu, and Yanxin Wang. "The effects of various LED (light emitting diode) lighting strategies on simultaneous biogas upgrading and biogas slurry nutrient reduction by using of microalgae Chlorella sp." *Energy* 106 (2016): 554–561.

Yusoff, Fatimah Md, Norio Nagao, Yuki Imaizumi, and Tatsuki Toda. "Bioreactor for microalgal cultivation systems: strategy and development." In *Prospects of Renewable Bioprocessing in Future Energy Systems*, pp. 117–159. Springer, Cham (2019).

Zhang, C. W., and A. Richmond. "Sustainable, high-yielding outdoor mass cultures of *Chaetoceros muelleri* var. *subsalsum* and *Isochrysis galbana* in vertical plate reactors." *Marine Biotechnology* 5, no. 3 (2003): 302–310.

Zhang, T. "Dynamics of fluid and light intensity in mechanically stirred photobioreactor." *Journal of Biotechnology* 168, no. 1 (2013): 107–116.

Zhang, Zhao, Jim Junhui Huang, Dongzhe Sun, Yuankun Lee, and Feng Chen. "Two-step cultivation for production of astaxanthin in *Chlorella zofingiensis* using a patented energy-free rotating floating photobioreactor (RFP)." *Bioresource Technology* 224 (2017): 515–522.

Zhu, Liandong, Zhongming Wang, Josu Takala, Erkki Hiltunen, Lei Qin, Zhongbin Xu, Xiaoxi Qin, and Zhenhong Yuan. "Scale-up potential of cultivating *Chlorella zofingiensis* in piggery wastewater for biodiesel production." *Bioresource Technology* 137 (2013): 318–325.

Zijffers, Jan-Willem F., Sina Salim, Marcel Janssen, Johannes Tramper, and René H. Wijffels. "Capturing sunlight into a photobioreactor: Ray tracing simulations of the propagation of light from capture to distribution into the reactor." *Chemical Engineering Journal* 145, no. 2 (2008): 316–327.

Zittelli, G. Chini, F. Lavista, A. Bastianini, L. Rodolfi, M. Vincenzini, and M. R. Tredici. "Production of eicosapentaenoic acid by *Nannochloropsis* sp. cultures in outdoor tubular photobioreactors." In *Progress in Industrial Microbiology*, vol. 35, pp. 299–312. Elsevier, 1999.

7 Bioelectricity Production Using Microbial Fuel Cell

7.1 INTRODUCTION

The Microbial Fuel Cells (MFCs) are biochemically catalyzed technologies that generate bioelectricity by oxidizing organic materials in the presence of electrochemically active microorganisms. The electron transfer to the anode by microorganisms via the oxidation of substrate creates an anodic half-cell potential in the anode chamber of MFC. Supplementation of synthetic redox mediators is required by mediator MFCs to make electron transfer from the bacterium to the anode more efficient. The electron transport may be aided by including a soluble mediator in the anolyte, which will shuttle the electron to the electrode surface and bacteria's redox enzyme(s). The capacity of the microorganisms to transmit electrons either directly or by self-synthesized electrons that shuttle from the cell to the electrode may be used to classify a mediator-less MFC. In mediator-less MFCs, microbes classified as "exoelectrogens," or electroactive bacteria, can transmit electrons directly to the anode.

7.2 CLASSIFICATIONS OF MICROBIAL FUEL CELLS (MFC)

7.2.1 Mediator MFC

There are two types of mediator MFCs:

- Indirect MFC—to generate potential in an anode it requires an external mediator for non-electrogenic bacteria.
- Mediator-driven MFC—Bacteria may produce their mediator, which can transmit electrons from the cell to the anode surface.

An oxidized mediator circulates in the anolyte of a mediator MFC. Then it reaches the cell membrane of the bacterium and is reduced by receiving electrons via bacterial oxidative metabolism. By giving an electron to an anode, the mediators are regenerated and continue to recycle via the redox process. Typically, such electron shuttles may receive electrons from different electron carriers inside the

DOI: 10.1201/9781003224587-7

cell (Figure 7.1). They subsequently vacate the cell by transferring the energy onto the electrode surface by crossing the cell membrane in reduced form (Logan, 2008).

Synthetic mediators are used in an anode chamber in an indirect mediator MFC which are typically redox molecules by nature. Reversible redox couples are formed by ubiquinone, dyes, and metal complexes (Rinaldi et al., 2008). Since they are stable in both reduced and oxidized states, these mediators were chosen as electron shuttle. They also do not disintegrate biologically and are not harmful to the microbial population. Unlike exoelectrogens or electrogenic microbes, microbes engaged in indirect MFC have different electron acceptors. They usually gather energy by providing electrons to oxygen, oxidized organic acid, and other compounds to complete the respiration cycle. The Gibb's free energy during the release of electrons is much more than the anode with the help of the above-mentioned electron acceptors. In normal settings, electrons cannot be transported from this microbial electron transport system to the anode because of the non-conductive nature of the cell surface structures. Therefore, these bacteria require some thermodynamically less favorable electron mediators to transfer electrons to the anode (Figure 7.2). To generate an anode half-cell potential synthetic or artificial mediator is used in an anode chamber with this kind of bacteria. To make electron transport from microbial cells to the electrode more efficient, electrochemical mediators are therefore engaged. The microorganism was used to investigate the efficacy of facilitated electron transfer from internal bacterial metabolites to the MFC anode using a variety of organic and organometallic compounds (Rabaey et al., 2007). The following are a list of organic dyes that were tested as electron mediators:

- phenothiazine
- phenoxazine
- indophenol
- phenazine
- thionine
- bipyridilium derivatives

It must be emphasized that the total efficacy of electron transmission mediators is influenced by several different factors. The re-oxidation of the mediator (having a specific electrochemical rate constant) depends on the electrode material. The above-mentioned artificial mediators have also termed electron shuttles. They increase the likelihood of producing reduced products with higher electrochemical activity than most microorganism-produced fermentation products. Furthermore, by infiltrating the outer cell membrane, these mediators may divert electrons from the respiratory chain. Then they are reduced, and they remain in this condition to shuttle electrons to the electrode. However, utilizing an artificial electron shuttle has the drawback of only recovering a portion of the electrons present in organic matter and accumulating end products in the anode chamber.

In some cases, microorganisms such as *Pseudomonas* sp., *Lactobacillus* sp., and *Enterococcus* sp., produce their mediators (phenazines, quinines, and flavin)

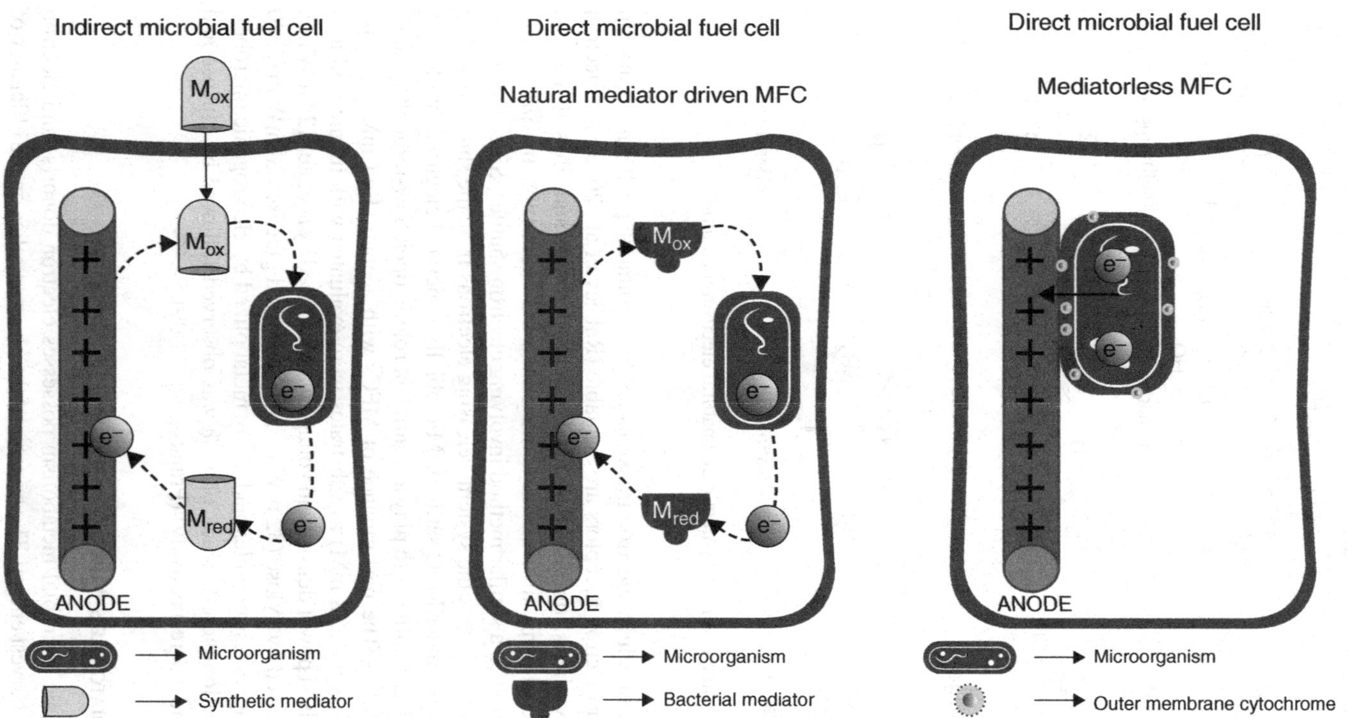

FIGURE. 7.1 Schematic representation of various electron transfer mechanisms in MFCs: a) a non-electrogenic fermentative microbe may be used in this indirect MFC. Further electrons which are available recover partially with external mediators; b) microbes may synthesize soluble mediators in a mediator-driven microbial fuel cell. by giving electrons to the anode, the mediators also get oxidized; c) microbes can attach to anode directly and transfer electron to it.

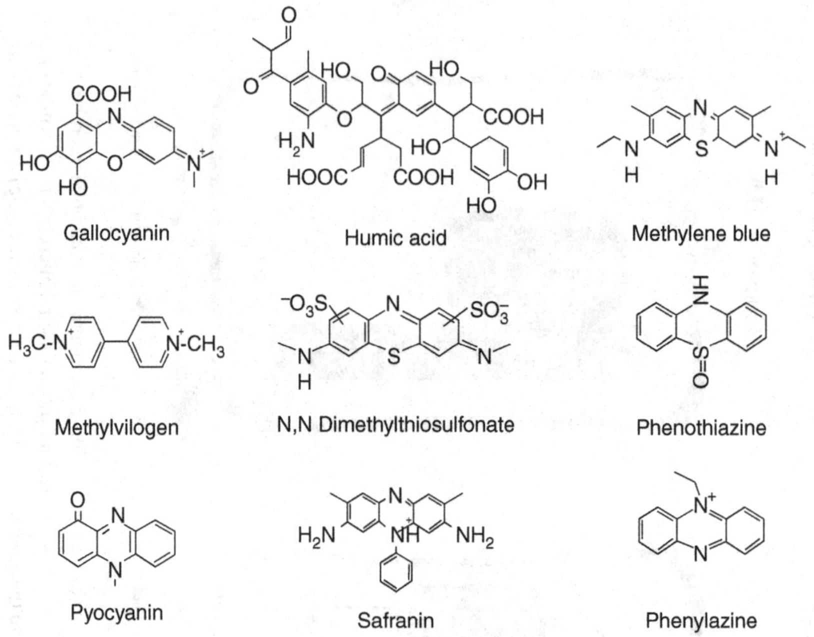

Gallocyanin Humic acid Methylene blue

Methylvilogen N,N Dimethylthiosulfonate Phenothiazine

Pyocyanin Safranin Phenylazine

FIGURE 7.2 Some common synthetic and natural electron mediators.

to promote extracellular electron transfer thereby maintaining cellular respiration when no other electron acceptors are available (Rabaey et al., 2010). An electron shuttle's biosynthesis, on the other hand, is energy inefficient, thus it must be recycled to regain this investment. Extracellular Fe^{3+} reduction by *Shewanella oneidensis* followed a similar method involving electron shuttles. Since the shuttle has not disappeared from the system, releasing electrons from the shuttle could be a good technique under MFC settings. Microbially generated electron shuttles, on the other hand, are unlikely to play a significant role in open flow-through systems (Kim et al., 2007). The performance of MFCs with acetate (1.64 mM), glucose (6.7 mM), and xylose (8 mM) as substrates were evaluated with humic acid as a mediator. Highest power density and voltage was produced by acetate (123 mW m^{-2}, 570 mV) followed by xylose (32 mW m^{-2}, 414 mV) and glucose (28 mW m^{-2}, 380 mV). As acetate is less complex and easily metabolized by electrogenic microbes, the highest coulombic efficiency of 14% was observed followed by glucose and xylose (8% and 5% respectively) (Franks et al., 2009).

7.2.2 Mediator-less MFCs

The electrogenic microbial metabolism possesses electron donor(s) and electron acceptor(s) to harvest energy. The transfer of the electron relies on the efficiency of the acceptors to accomplish their respiration. A large number of bacteria donates their electron to metal ions (oxidized state; such as Fe^{+3}, Cr^{+6}, Mn^{+4}) and sulfate ion (SO_{4-}^{2}) in a natural condition to respire. To carry out their respiration in MFC, most of them tend to spontaneously release electrons to the anode as the final electron acceptor. In mediator-less MFC, continuous donation of an electron by these exoelectrogens such as *Geobacter* and *Shewanella* species are responsible for developing anode potential. Due to the lack of a natural electron acceptor, the electron emitted by exoelectrogens is transported to the anode. Biochemical and genetic characterizations of exogenous electron transfer showed that the outer-membrane cytochromes are important in establishing the relay of electrons. Furthermore, some bacteria form and utilize soluble electron shuttles, which minimize the necessity for direct interaction between the cell and the electron acceptor. Phenazine synthesis by *Pseudomonas aeruginosa*, for instance, restored electron transport for several bacterial strains (Rabaey et al., 2005).

7.3 MICROBIOLOGY AND BIOCHEMISTRY OF MICROBIAL FUEL CELLS

7.3.1 Microbiology of Electrogens

Electrogenic microorganisms are the biocatalyst that play a vital role in MFC. These microorganisms are responsible for donating electrons and developing negative potential on the anode. Since the current generation adds to anodic internal resistance, the overall energy density is affected by the inoculum type or microbial culture. Microorganisms from both mixed and pure cultures were included as inoculum in MFC. Initially, studies on MFC started with the use of non-electrogenic microorganisms in anode along with synthetic mediators (Behera and Ghangrekar, 2009). In this study, sludge has been fed to a dual-chamber MFC and there was a peak power density of 600 mW m^{-2}. In another study, methylene blue, 2-hydroxy-1,4-naphthoquinone, and neutral red as mediator was used along with anaerobic sludge as inoculum (Taskan et al., 2015). On using methylene blue, there was a peak power density of 600 mW m^{-2}. In all these above studies, the inoculum was sourced from an anaerobic digestor which comprises consortia of different types of microbes.

The use of *Shewanella putrefaciens* as anodic inoculum in MFC was the first breakthrough in terms of using defined and specialized electrogenic bacteria for bioelectricity generation. The anode may directly receive electrons from this kind of bacterium, which was demonstrated by Kim et al. (2002). Various pure cultures of electrogenic microbes were exploited in the anode of an MFC. The *Geobacter* sp. (Bond and Lovley, 2003), *Shewanella* sp. (Kim et al., 2002), *Rhodoferax* sp. are a few among them (Liu et al., 2017). In certain cases, the inoculum was a mixed culture. The following are the sources from which we gathered our mixed culture:

- domestic wastewater (Min and Logan, 2004)
- fresh and marine sediments (Zhang et al., 2009)

- soil (Niessen et al., 2004)
- activated sludge (Ki et al., 2008)
- anaerobic digester sludge (Chae, 2010)

7.3.1.1 Axenic Electrogenic Microbial Cultures

There are few bottlenecks in experimenting with pure culture systems in MFC. When pure culture is used in MFC, there are immense chances of microbiological contamination with non-desired properties. In contrast to mixed-culture systems, pure culture can only use a few types of substrates (Sharma and Kundu, 2010). Pure exoelectrogenic cultures have been used as MFC inoculum and have produced significant power densities (Lovley and Nevin, 2008). For improving the robustness of MFCs, the use of mixed cultures like anaerobic digester sludge has been explored as they are widely accessible in big quantities and are more resistant to environmental changes.

7.3.1.2 Exoelectrogens Isolated from MFC

These were extracted directly from MFCs in large numbers. Microorganisms linked with electrodes in MFC systems include a wide range of species. A common term used for bacterial mat over the surface of the electrode is biofilm. Implementation of BioLP (biological laser printing) to isolate pure electrogenic bacteria has been explored recently (Ringeisen et al., 2010). The *Bacteroides* sp. W7, a Fe (III)-reducing fermentative bacterium was isolated from anodic solution (Wang et al., 2010). After enriching an exoelectrogenic *Ochrobactrum anthropi* YZ-1 was isolated using dilution to extinction procedures in a U-Tube MFC (Zuo et al., 2008). A new Fe (III)-reducing bacteria was discovered using the acetate enrichment method that was electrochemically active and was phylogenetically related to *Aeromonas hydrophila* (Pham et al., 2003). Very few obligate anaerobes were reported for MFC application. The *Clostridium butyricum* EG3, an obligate anaerobe, was isolated from a mediator-free MFC employing starch processed wastewater as an anode in one investigation (Park, 2001). It is vital to remember that not all of the organisms in the biofilm will interact with the anode directly. They may, however, have an indirect interaction via other members of the electrogenic microbial community (Park et al., 2001). For example, *Brevibacillus* sp., PTH1 power output was found to be minimal until it is co-cultured with a *Pseudomonas* sp. Moreover, if MFC was operated with only *Pseudomonas* sp. then also the power generation was found to be suboptimal (Boon et al., 2008).

In mixed consortia MFCs, the role of diverse species is unclear. Non-current producing microorganisms may impair or assist the formation of biofilm on the anode, as well as compete with the planktonic culture. Thus, the isolation procedure must be correctly chosen for selective beneficiation and isolation of electrochemically active bacteria from an anode surface of MFC. Pure cultures would be ideal for research purposes at a laboratory scale. But when it comes to practical and industrial applications, mixed culture may hold the upper hand in terms of robustness and maintenance of an electrogenic environment.

7.3.1.3 Mixed-culture-based MFCs

The startup time for an MFC is crucial as it sets the tone of electricity generation and the rate of substrate degradation. The shorter the startup time, the lesser the time taken for attaining optimum performance of MFCs. It has been observed that the startup time taken by pure-culture-based MFCs is shorter when compared with a mixed culture. Furthermore, the mixed culture takes more time to reach a consistent power output. The startup time of the MFC system may be described as the time it takes to choose and strengthen its electrochemically active microbial consortia from mixed culture under suitable acclimatization conditions (Kim et al., 2007). The MFCs have a broad variety of startup times, ranging from 4 to 103 days. The majority of the microbes in these consortia belonged to γ-proteobacteria and α-proteobacteria. The microbial communities get influenced by the operational conditions, nature of the carbon source, and reactor configurations (i.e., anolyte salinity, external resistance). Mixed culture consists of vivid profile viz. *Saccharomyces sp.*, *Proteus vulgaris*, *Escherichia coli*, *Clostridium butyricum*, *Clostridium beijerinckii*, *Saccharomyces cerevisiae*, *Rhodoferax ferrireducens*, *Pseudomonas mendocina*, *Gluconobacter oxydans*, *Stenotrophomonas acidaminiphila*, *Pseudomonas pseudoalcaligenes*, *Bacillus subtilis*, *Paenibacillus lautus*, *Alcaligenes faecalis* and *Enterococcus gallinarum*.

In one such report, when MFC was fed with wastewater having sulfate intermediates, 70–80% of the power generated was due to the activity of sulfur-reducing bacteria (*Desulfovibrio* sp, *Geobacter* sp.,) whereas only 20–30% was due to exoelectrogen (Ieropoulos et al., 2005). A current density of 12.30 ± 0.01 A/m^2 and a very high power density of 4.99±0.02 W/m^2 were observed with a mixed culture dominated by *Geobacter* sp. and fed with starchy wastewater (Wang et al., 2017).

7.4 BIOCHEMISTRY OF MICROBIAL FUEL CELL

7.4.1 Biochemistry of the Mediator-based Electron Transfer Methods

The performance of MFCs operating with microbes that cannot shuttle their electron to the anode can be improved remarkably if electron-shuttling mediators are added externally. Electrons may be transferred between bacteria and electrodes using low molecular weight redox species. Thionine, methyl vilogen, safranin, methylene blue, indophenol, and humic acid are a few mediators which are used to shuttle electrons in MFCs. However, mediators must perform the following responsibilities to offer an effective electron transfer pathway from bacterial metabolites to anode:

- The mediator's oxidized state can quickly penetrate the bacterial membrane and approach the reductive species within the bacterium.
- The mediator potential must be high enough to allow for rapid electron transport from the metabolite, but not so high that it prevents considerable potential loss.

- The oxidation state of the mediator must prevent it from being degraded by other metabolic processes.
- The decreased form of the mediator must readily pass through the bacterial membrane.
- In the electrolyte solution, both oxidation states of the mediator must be chemically stable.
- The mediator must be water-soluble and not adhere to the electrode surface or bacterial cells.

There are three different ways to couple up the mediators with the microorganisms.

- A covalently connected diffusional mediator that shuttles between the microbial cells and anode.
- A diffusional mediator connects the microbial suspension to the anode surface.
- The mediator binds to microbial cells and aids in transferring electrons from the cells to anode.

7.4.1.1 Biochemistry of Indigenous Mediator Producing Electrogenic Microbes

To promote extracellular electron transfer, some specialized microbes can produce their mediators. Electron transfer to Fe^{3+} was initially reported in *Shewanella oneidensis* and thereafter in organisms like *Pseudomonas* and *Geothrix fermentans* sp. An electron shuttle should be recycled several times to make use of this energy investment since biosynthesizing an electron shuttle is energy intensive. In open and dynamic environments, microorganisms that produce electron shuttles rapidly lose their availability due to dilution or relaying electrons to nonessential Intermediates. The microbial fuel cell isolates *Pseudomonas putrefaciens* and *Pseudomonas aeruginosa* developed phenazine as electron shuttles that might help in electron transport to electrodes (Rabaey et al., 2005). Furthermore, investigations of microbial profiles using 16S rRNA gene sequences revealed the existence of *Enterococcus* and *Lactobacillus* species, both of which might emit electrochemically active compounds (Rabaey et al. 2004). It is possible to release an electron shuttle under these circumstances since the shuttle is not lost in the system. It was observed that when the components of the anodic chamber of *Geothrix*-based MFCs were replaced entirely with fresh media, the power output decreased by 50% (Bond and Lovley, 2005). There is a net loss of energy to the organism if an electron shuttle is flushed out of the system repeatedly.

7.4.2 Electron Transfer in Mediator-less MFC

Anode respiration is a term used to describe bacteria in mediator-less MFC. They do not need any mediators to transmit electrons to the anode in the respiration process. Exogenous electron transport may include the outer-membrane cytochrome, according to biochemical and genetic characterizations. Microorganisms such as *Geobacter, Clostridium butyrium, Aeromonas hydrophila, Shewanellaputrfaciens,* and *Pseudomonas aeruginosa, Rhodoferax ferrireducens, Shewanella oneidensis,* and have all been shown to oxidize organic. This helps in the termination of their metabolic process.

Power production in MFC was confined to bacteria capable of dissimilatory iron reduction, according to the fundamental results. As a result, it was shown that MFCs included a wide range of microbial communities. The ability to produce current has been shown in four of the five Proteobacteria classes, and the Acidobacteria and Firmicutes phyla. Redox enzymes were found on the outer membrane of the yeast *Pichia anomala,* which aided current generation in an MFC (Prasad, 2007). The *Synechocystis* sp. PCC 6803, an oxygenic phototrophic cyanobacterium, was studied to see whether it could form electrically conductive appendages termed nanowires (Gorby, 2008). Pili generated by certain bacteria are proven to be electrically conductive with scanning tunneling electron microscopy. An additional study is needed, however, to fully understand the function of the nanowire in active electron transport.

The most essential process is the direct transmission of electrons from microbes to electrodes. When *Shewanella putrefaciens* cultures generated power during metabolizing lactate, it was originally assumed that microorganisms would be able to transmit electrons directly to an electrode surface. The metal-reducing bacteria *Shewanella putrefaciens* MR-1 is claimed to contain cytochromes in its external membrane. Such electron carriers (i.e., cytochromes) can produce anodic current in anaerobic conditions in the absence of terminal electron acceptors. Cytochromes found on the outer membranes of *Shewanella* have a function in electron transfer reduction.

It is possible to use two forms of direct transfer of electrons from the microorganism to the electrode:

- Exoelectrogen mediates the electron transport between electrode and bacteria through endogenous mediator secretion.
- After the bacteria have formed a biofilm on the electrode, direct transmission of an electron to the electrode through pili or "nanowire" occurs.

Metal-reducing bacteria have been shown to have similar mechanisms of electron transfer as MFCs, despite the lack of precise information on the MFC process. A solid substrate (metal) outside of the bacterium accepts the last electrons of substrate metabolism here.

Only by combining biochemistry and electrochemistry can the electron transmission pathway from the microbe to the anode be evaluated. Regardless of the method, the electrical potential between the final carrier and the anode determined the extracellular electron-transfer rate. Because electrons need to be transferred from a high-energy state (the highest negative potential) to a lower-energy state (the lowest negative potential). As a result, a large potential difference in the MFC operation will allow for a greater electron discharge. The flow of electrons in MFCs creates a redox atmosphere that may be precisely regulated. It is necessary to understand how microorganisms collect and utilize energy to comprehend how an MFC generates electricity. Bacteria proliferate by activating chemical processes and storage of energy in the ATP form ("adenosine triphosphate"). Reduced substrates are oxidized in a few bacteria, and NADH transfers electrons to respiratory enzymes. This is the nicotinamide adenine dinucleotide (NADH) in its simplest form (NAD). A respiratory chain transports these electrons downward. Protons are transported across an internal membrane by a sequence of enzymes, resulting in a proton gradient. The enzyme ATPase transports the protons back into the cell. For every 3–4 protons, 1 adenosine diphosphate produces 1 ATP molecule. The electrons are then released into a soluble terminal electron acceptor including sulfate, oxygen, nitrate, or an electrode (in a bio-electrochemical system). The maximal potential of method is ~1.2 V under standard conditions, depending on the potential difference between oxygen and electron carrier (NADH) (Rabaey, 2008).

7.5 METHODS FOR UNDERSTANDING PERFORMANCE OF MFC

The performance of an MFC depends on several physical, chemical, and biological factors. High power production and strong electron recovery efficiencies are required for MFCs to be accepted as a competitive technology for renewable energy sources. MFCs may suffer power loss in different ways:

- Activation loss (the processes on both electrodes and the extracellular electron transmission to the anode are likely to be started).
- Bacterial metabolism loss (by using oxidation to get their energy from their substrates).
- Mass transfer loss (because of the limited flow of the reactants to the electrode).
- Ohmic losses (on account of the resistance in the proton diffusion and transfer).

The MFCs always produce low operating voltage (V_{op}) compared to the cell electromotive force (E_{thermo}) (Figure 7.3). The irreversible losses of operating voltage in a MFC can be explained by Eq. 7.1 (Rismani-Yazdi et al., 2007).

$$V_{op} = E_{thermo} - \left[(\eta_{act} + \eta_{ohmic} + \eta_{conc})_{\text{cathode}} + (\eta_{act} + \eta_{ohmic} + \eta_{conc})_{\text{anode}} \right] \quad (7.1)$$

ohmic loss from ionic and electronic resistances is denoted by η_{ohmic}, and the concentration loss due to mass transport limitations by η_{conc}, activation loss because of reaction kinetics by η_{act}. It is also reported that the performance of MFCs is collectively limited by both cathode and anode overpotentials. The electrodes' over-potentials are generally influenced by current, which may be signified by a polarization curve. For the investigation and characterization of fuel cells, a

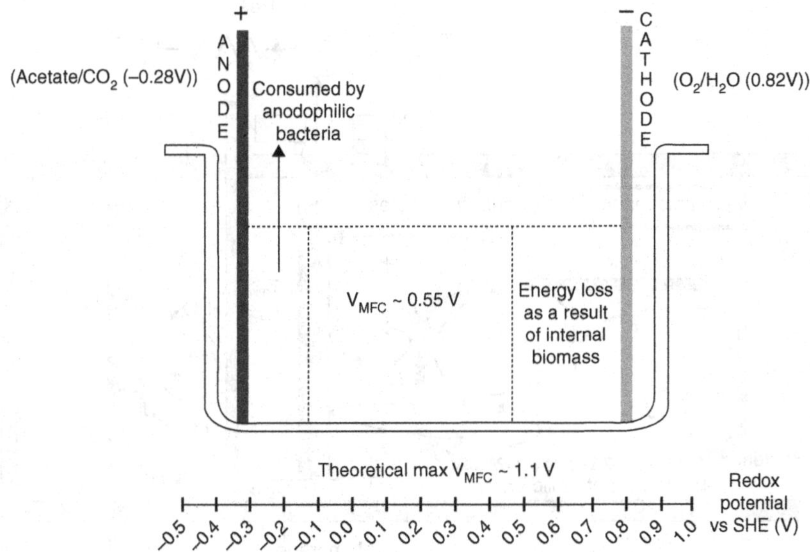

FIGURE 7.3 Voltage generation in MFC.

polarization curve is a useful tool. It is demonstrated that voltage changes as a function of the current density. Three types of potential losses occur in a MFC viz. activation losses (due to metabolic eccentricities of electrogenic bacteria), ohmic losses (due to internal resistance) and concentration polarization (due to mass transfer) (Figure 7.4).

7.5.1 Activation Overpotential

The anode and cathode, correspondingly, need activation energy to initiate the oxidation and reduction processes. Activation overpotentials occur during the startup phase of electron transport from (or) to a chemical reacting at the electrode surface. The activation energy necessary to either oxidize or reduce a chemical on the anode surface is represented by this potential at the anode. On the other side, this potential at the cathode is the activation energy barrier. The oxidant's conversion into a reduced state is hampered as a result. This also restrains the reduction kinetics. At low current densities, the activation losses cause an exponential loss on the current-voltage curve. As additional current is extracted from MFC, the losses rise once again. This in turn results in a lower operating voltage of MFC. At lower current density the activation losses are dominant. Adsorption of reactant species and desorption of product species on the electrode surface are two of the most important elements in activation overpotential (Rabaey et al., 2009).

Various tactics were used to decrease the activation overpotential. Increased electrode catalysis and the addition of mediators to enable effective electron

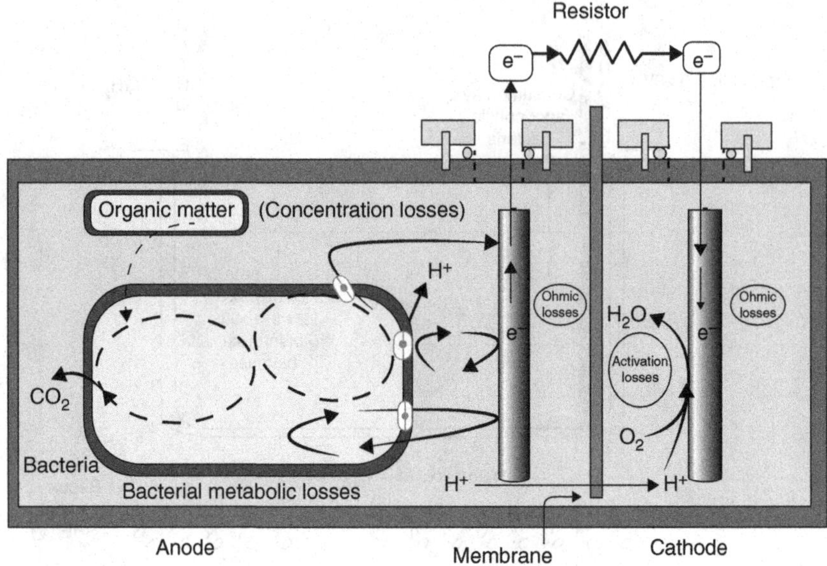

FIGURE 7.4 Types of over-potentials observed in MFCs: a) activation losses, b) ohmic losses, c) concentration polarization.

transmission from the bacterium cell to the anode surface may help to reduce it. It may also be reduced by expanding the electrode surface area and enriching the electrogenic biofilm on the anode. An operating condition within cathode and anode sections also helps in minimizing the activation overpotential (Rismani-Yazdi et al., 2008). Activation polarization is reduced in mediator-free MFCs due to conducting pili. Activation polarization is also faced by a cathodic reaction. Platinum (Pt) is chosen as a cathode for oxygen reduction reactions over non-catalyzed graphite. Because the cathodic oxygen reaction that generates water has a lower energy barrier, it is more efficient.

7.5.2 Ohmic Overpotential

Ohmic overpotential is caused by the resistance to ions flow in electrolytes. Also, it is caused by the electron flow between the anode and cathode. By reducing electrode spacing the ohmic loss in electrolytes can be reduced as it is dominant. The current production is proportional to the amount of ohmic overpotential in the system. For reduction, this will ensure counter-ion migration and proton availability on the cathode surface.

Another way to lower ohmic overpotential is to increase the electrolyte's ionic conductivity. The presence of a separator in an MFC prevents ions from migrating through the electrolyte. This transmembrane resistance is a major contributor to ohmic losses. This may be reduced by employing high conductivity membranes.

The improvement of the membrane surface region improved overall MFC efficiency by improving the proton flow and minimizing ohmic losses (Du, 2007). Moreover, biofouling on the membrane rises the transmembrane resistance. In one study, effect of fouling on anion exchange membrane (AEM), cation exchange membrane (CEM) and Nafion (proton exchange) membrane on the performance of MFC was investigated. The current generation drastically dropped by 60%, 42% and 47% on using cation exchange membrane, anion exchange membrane and proton exchange membrane, respectively (Ping et al., 2013). The decrease in current could be due to membrane fouling. The CEM are prone to inorganic scaling-based fouling whereas AEM are prone to biofouling.

7.5.3 Concentration Overpotential

Concentration overpotential in MFC may be attributed to the slow mass transfer rates of reactants and products inside the reactor. At high current densities, concentration loss occurs. With an increase in current, the voltage drops rapidly. Due to mass transfer limitations, the substrate is sparsely available for the biofilm on the anode. As a result, concentration losses occur (Logan et al., 2008). Similarly, the unavailability of dissolved oxygen (DO) in the cathode chamber is responsible for cathodic overpotential. As a result, the power density output of certain MFCs is limited. To minimize the concentration overpotential, homogenous turbulent mixing electrolytes is one way. The concentration gradient in an MFC can be reduced by stirring and/or bubbling. External energy input is needed for this kind of mechanical operation. The power requirement is generally greater than the power output obtained from the MFC. In turn, it makes the MFC operation expensive and unsustainable. Concentration overpotential may be decreased by raising electron donor concentration in the anode compartment and electron acceptor or proton concentration in the cathode compartment. The growth of biofilm on an anode may be inhibited by high concentration, so substrate concentration in the anode requires optimization. In a study, three-fold improvement in current density was observed when the permanganate concentration (electrolyte) as the cathodic electron acceptor was raised between 0.02 to 0.2g L^{-1} (You et al., 2006). In the functioning of a microbial fuel cell (MFC), power overshoot may occur when the electrical current is high. This may happen when external loads are reduced and both cell voltage and electrical current drop.

7.5.4 Effect of Cathode Performance

The cathode's performance is important in overall power production in MFC. The final electron acceptor concentration decreases in cathodic chamber due to capturing of electrons and protons produced at the anode. The electrons flow through a resistor and reach the cathode, where they are captured by catholyte thereby producing electricity. To maintain electro-neutrality, protons flow from the anode to the cathode through an ion exchange membrane. Oxygen is a common option as electron acceptor due to its limitless abundance and high standard redox potential. There are two approaches to reducing oxygen at the cathode:

- Water generation via a four-electron pathway.

$$O_2 + 4H^+ + 4e^- \rightarrow 2H_2O \ (E^0 = 0.816V) \tag{7.2}$$

- Inadequate oxygen reduction resulted in the creation of hydrogen peroxide, which has a poor energy conversion efficiency.

$$2O_2 + 4H^+ + 4e^- \rightarrow 2H_2O_2 \ (E^0 = 0.295V) \tag{7.3}$$

The voltage output of the MFC was generally found to be smaller than the thermodynamic ideal voltage in most of the circumstances. This occurs as a result of irreversible voltage losses or overpotential (Freguia et al., 2007). Activation losses, mass transport losses, as well as ohmic losses, all have an impact on MFC performance. In MFC, a large portion of the produced voltage is needed to compensate for current losses in the cathode because of electrochemical reactions, mass transfer processes, and charge transport.

The kinetics of the reduction at the cathode determines how much power would be produced in MFCs. The kinetics of the process is slowed by an activation energy barrier that prevents oxygen from being converted to water. Because of this activation barrier, cathode potential is being lost during current MFC generation (η_{cat}, cathodic activation loss). The voltage drop needed to power the electron and proton transport pathways is represented by the ohmic overpotential. The electronic (R_{elec}) and ionic (R_{ion}) resistances combinedly contributes towards cathodic ohmic resistance. This comprises electrolytes, electrode resistance, and interconnections:

$$\eta_{ohmic} = iR_{ohmic} \tag{7.4}$$

$$R_{ohmic} = R_{ion} + R_{elec} \tag{7.5}$$

Transport losses affect the availability of oxidants and the elimination of products. The reaction rates, as well as Nernstian cell voltage, are both affected by reactant consumption or insufficient oxidant resulting in suboptimal performance. At high current densities, mass transport losses occur, and the amount of these losses grow with increase in current density. In comparison to the anode, the cathode gets significantly influenced by mass transport loss.

Cathode performance is affected by the following important factors:

- Cathode surface area
- Cathode material
- Cathodic electron acceptor
- Cathode catalyst
- Cathodic operational conditions including catholyte pH, cathode oxidant concentration (Wei et al., 2011)

7.6 MICROBIAL FUEL CELL DESIGN

Apart from component-by-component optimization, an extensive study was conducted to determine the best MFC design. The MFC's coulombic efficiency and power production are largely determined by reactor architecture. Researchers first employed dual-chambered MFCs, but single-chambered MFCs were subsequently used to improve the volumetric power density. To maximize power generation, stackable MFCs were tested in series and parallel configurations.

7.6.1 Dual-chambered MFC

A cathode chamber and an anode chamber are combined in a dual-chambered MFC. A separator connects the two chambers. Ions may move from anode to cathode chamber through the separator. Neither the cathode chamber nor the anode chamber can receive oxygen or other oxidants from the other compartment. It is possible to run the MFC in batch or continuous mode and with different compartment shapes. The H-type MFC was a frequently utilized dual-chambered MFC (Mohan et al., 2008); however, the H-type reactor's power production is restricted by its high internal resistance (Rinaldi, 2008).The MFCs with dual-chambered up-flow modes have been proven to be effective for wastewater treatment. Pumping fluid around, on the other hand, requires a lot more energy in comparison to their power production. In one study, a 1.2 mL dual-chambered MFC with a 2 cm^2 cross-sectional area by using this design. A 500 W m^{-3} was the highest volumetric power density achieved due to large electrode surface area to anolyte volume ratio and most modest electrode spacing (Ringeisen et al., 2007). In a single electrode/PEM assembly in FPMFC (flat plate MFC), the anodic chamber, like a conventional fuel cell, would be supplied with a substrate, and the cathodic chamber would be pumped with dry air instead of liquid catholyte (Min and Logan, 2004). Both processes can be done in a continuous-flow mode. Due to their complicated design, high internal resistance, and vast volume, dual-compartment MFCs are difficult to scale up.

7.6.2 Single-chambered MFC

The single chamber MFCs usually just have an anodic chamber, since a cathodic chamber is not required for aeration. Since the cathode is in direct contact with the air, the complexity of designing a single-chambered MFC is much lesser when compared to dual chamber. A MFC with a rectangular anode linked to a porous air cathode that is subjected to the air directly showed greater efficiency (current density of 788 W m^{-2}) as compared to dual chamber (Park and Zeikus, 2003). The advantages of single-compartment MFC are mentioned below:

- As passive air is utilized as catholyte, there is no extra power consumption required for aeration in the cathode chamber.
- It is compact and easy to scale up.

- Requires less effort because catholyte does not need to be recycled or chemically regenerated.
- Reduced cell capacity; greater volumetric power density due to short electrode spacing configuration.

Membrane-less air cathode MFCs offer many benefits, including a cheap cost, a basic arrangement, and a reasonably high power density with a small reactor capacity. However, as compared to a single-compartment MFC with a membrane, they have a poor coulombic efficiency (CE). The low CE achieved in membrane-less MFC is due to the fact that anode chamber getting exposed to oxygen (which act as an electron acceptor) when electrons was generated by substrate metabolism (Liu and Logan, 2004). Also, the possibility of a short circuit due fewer electrode spacing (about 1–2 cm) limits the use of membrane-less MFC.

In a new innovative MFC design, a carbon cloth electrode assembly was developed by sandwiching a J-cloth between cathode and anode. This greatly lowered the internal resistance and resulted in a power density of 627 W m^{-3} in fed-batch mode and 1010 W m^{-3} in a continuous-flow mode (Fan et al., 2007). This is the highest reported power density for MFCs and more than fifteen times higher than those reported for air-cathode MFCs using similar electrode materials.

7.6.3 Stacked MFC

In recent years, researchers have tried a variety of engineering techniques to improve MFC performance. This improvement can mainly be attributed to relieving physical and chemical constraints through electrode material and reactor architecture improvement and optimization of operational conditions. However, due to thermodynamic restrictions, low output current and low operating voltage, the electrical power produced by a single MFC cell cannot be directly used as a reliable power source. Moreover, due to the various factors mentioned in the above sections, the output power obtained is not continuous and fluctuates. An increase in cell volume would not proportionally increase power generation due to the domination of internal losses. Through the use of a series connections between many fuel cells, voltages may be increased while still maintaining a constant current (Barbir, 2012). Fuel cells coupled in parallel increase current output and average voltage. Fuel cells or other power sources may be coupled in series or parallel to achieve any required current or voltage. The series connection produces six times less current than the parallel connection, suggesting that connecting MFCs in series does not allow for high current densities (Aelterman et al., 2006). Individual MFCs or MFCs connected in parallel were studied for rapid substrate degradation and high current densities. It was observed that a voltage reversal issue may occur when MFCs are connected in series leading to poot voltage output. Poor efficiency of individual MFCs in this series connection might limit the amount of current generated. Moreover, the bacterial activity gets hampered due to the voltage reversal problem. A parallelly connected MFCs can negate the above-mentioned problem and has emerged as an alternative stack configuration. However, significant challenges remain in harvesting electrical power from multiple MFCs operating at numerous voltage and current levels. It is reported that larger power production cannot be easily achieved by building larger MFCs or simply connecting them in series or parallel because of the nonlinear nature of MFCs.

7.7 PROTOCOL FOR OPERATION OF A DUAL-CHAMBER MFC

Study of bioelectricity production along with the wastewater treatment through a batch process and the subsequent analysis of current density, coulombic efficiency, and power density can be done using dual-chamber MFCs. This experiment uses *Shewanella* sp., an obligate anaerobe, and exoelectrogen, in which electrons flow through exocytosis and produce ATP via an electron transport chain of oxidative phosphorylation. An aqueous solution or solid conductor contains the final electron acceptor, a strong oxidizing agent.

Materials required:

A) Instruments: Digital multimeter, double-chambered MFC reactor, Ag/AgCl –KCl (+197 mV) reference electrode, data acquisition system, test tubes, resistance box, pH meter, COD designer, Spectrophotometer.
B) Medium:
  - Synthetic acetate-based wastewater (initial COD = 3000 mg/L)

• CH_3COONa	3.843 g L^{-1}
• NH_4Cl	0.954 g L^{-1}
• $NaHCO_3$	4.5g L^{-1}
• K_2HPO_4	0.081 g L^{-1}
• KH_2PO_4	0.27 g L^{-1}
• H_3BO_3	0.106 mg L^{-1}
• $CoCl_2.6H_2O$	52.6 µg L^{-1}
• $(NH_4)_6Mo_7O_{24}.4H_2O$	52.6 µg L^{-1}
• $CuSO_4$. $5H_2O$	4.50 µg L^{-1}
• $ZnSO_4$. $7H_2O$	0.016 mg L^{-1}
• $MnSO_4$. H_2O	0.526 mg L^{-1}
• $NiSO_4$. $6H_2O$	0.526 mg L^{-1}
• $FeSO_4$. $7H_2O$	10 mg L^{-1}
• $MgSO_4$. $7H_2O$	0.06 g L^{-1}
• $CaCl_2$. $2H_2O$	0.75 g L^{-1}
• pH	7.0

7.7.1 PROTOCOL OF ASSEMBLY OF MFC

The dual-chamber MFC is assembled with graphite electrode in both the chamber. The synthetic wastewater is poured in the anode chamber. The anode chamber is

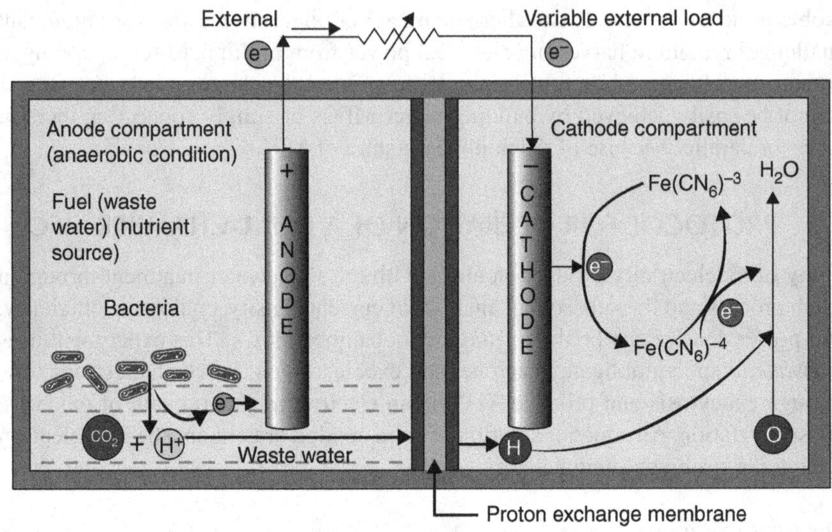

FIGURE 7.5 Schematic representation of a dual-chamber MFC.

then sparged with nitrogen gas to create anaerobic condition. Aseptically, active culture of *Shewanella* sp., is injected in the anode chamber. The microbe was allowed to form biofilm over the anode (takes one week). In the cathode chamber iron (III) hexacyanoferrate (1 g L^{-1}) solution was used as catholyte (Figure 7.5).

a) Polarization study
   - Pretreated overnight grown inoculum of 10% v/v is mixed and allowed to ferment via the batch process for nearly 36 hours with the addition of N_2 gas to achieve an anaerobic environment in the reactor.
   - At the steady-state of voltage generated in the closed-circuit system, the polarization curve is plotted with the different values of the external resistance box (90,000 Ω to1 Ω) stepwise and the voltage drop in each case is monitored by using a multimeter. After 3–7 minutes of stable data readings, the circuit is reused to take new readings.
   - Whenever the process is completed, 5 ml of broth is collected from the reactor for the determination of coulombic efficiency followed by COD removal.

b) Cyclic voltammetry (CV) study

By monitoring the current response at an electrode surface to a specified range of applied potentials, cyclic voltammetry (CV) is an important approach for determining the mechanisms of electrode reactions underpinning oxidation or reduction reactions on the electrode surface. Peaks in the voltammogram indicate an increase in current when the voltage at the electrode surface reaches a species' standard reduction or oxidation potential (current vs. applied voltage curve). Electrons will be moved to or from the electrode surface depending on whether the electrode is being reduced or oxidized. To get reliable results, a three-electrode setup (working electrode, reference electrode, and counter electrode) is required. The parameters of CV are determined by a number of elements, including electrode surface pre-treatment, electron transfer reaction rate, chemical and biological species present and their thermodynamic properties, electroactive species concentration and diffusion rates, and potential scan rate.

The CV experiments were conducted extensively in MFCs to, a) elucidate the exocellular electron transfer mechanisms of anode reactions involving both direct and indirect electron transfer between the biofilm and the electrode, b) determine the redox potentials of the chemical or biological species involved at the anode, and c) evaluate the performance of the catalysts being studied for anode surface modification (Zhao et al., 2009). Forward and backward voltage sweeps with rates in the range of 1–20 mV s^{-1} are commonly used in MFC investigations that use CV. Multiple peaks in the cyclic voltammograms may be noticed due to the existence of several distinct redox species in MFC (Harnisch and Freguia, 2012).

c) Calculations:

Open circuit voltage (OCV) which is observed before attaching resistance, can be determined using multimeter. Once an external resistance (R) is introduced to close the circuit, the current is generated (I).

$$\mathrm{I} = \frac{OCV}{R}\,(\text{Ampere}) \tag{7.6}$$

$$\text{Power (P) can be calculated as: } \mathrm{P} = \mathrm{I} \times \mathrm{V}\,(\mathrm{W}) \tag{7.7}$$

$$\text{Current density (CD) can be calculated as: } CD = \frac{Power}{V_{an} \times R}\,(A\,m^{-2}) \tag{7.8}$$

where, V_{An} is volume of the anode.

$$\begin{aligned}\text{Power density (PD) can be calculated as: } & PD = CD \times OCVPD \\ = CD \times OCVPD = CD \times OCVPD = CD \times OCV\,(\mathrm{W\,m^{-3}})\end{aligned} \tag{7.9}$$

Calculation of coulombic efficiency (C_E):

The percentage of substrate used to generate current is known as columbic efficiency. CE is determined as a ratio of total recovered coulombs to theoretical coulombs that can be produced from the substrate by integrating the current over time.

$$C_E \frac{M\int_0^t b\ I\ dt}{F\ b\ v\ \Delta COD} \tag{7.10}$$

Where, $\int_0^{tb} I\,dt$ is the current measured at time t_b, ΔCOD is the change in COD, v = working volume of the anode chamber and t_b = the batch time in second.

CONCLUSIONS

MFCs have come a long way as a technology in the previous decade, with significant advancements. Increases in current and power density, materials (electrodes and membranes), design, size, and the diversity of biocatalysts used have all contributed to these advancements. In terms of substrates, several types of wastewaters have been investigated as MFC substrates over the years. Initially, the attention was mostly on substrates that contained a mixture of carbon sources, such as acetate and glucose. Focus on more sophisticated anode and cathode electrodes, specific MFC designs, and improvements in power management could increase power density and potentially broaden MFC uses as biological sensors or in sensor networks.

REFERENCES

Aelterman, Peter, Korneel Rabaey, Peter Clauwaert, and Willy Verstraete. "Microbial fuel cells for wastewater treatment." *Water Science and Technology* 54, no. 8 (2006): 9–15.

Barbir, Frano. *PEM Fuel Cells: Theory and Practice*. Academic press, 2012.

Behera, Manaswini, and M. áM Ghangrekar. "Performance of microbial fuel cell in response to change in sludge loading rate at different anodic feed pH." *Bioresource Technology* 100, no. 21 (2009): 5114–5121.

Bond, Daniel R., and Derek R. Lovley. "Electricity production by *Geobacter sulfurreducens* attached to electrodes." *Applied and Environmental Microbiology* 69, no. 3 (2003): 1548–1555.

Bond, Daniel R., and Derek R. Lovley. "Evidence for involvement of an electron shuttle in electricity generation by *Geothrix fermentans*." *Applied and Environmental Microbiology* 71, no. 4 (2005): 2186–2189.

Boon, Nico, Katrien De Maeyer, Monica Höfte, Korneel Rabaey, and Willy Verstraete. "Use of *Pseudomonas* species producing phenazine-based metabolites in the anodes of microbial fuel cells to improve electricity generation." *Applied Microbiology and Biotechnology* 80, no. 6 (2008): 985–993.

Chae, Kyu-Jung, Mi-Jin Choi, Kyoung-Yeol Kim, F. F. Ajayi, In-Seop Chang, and In S. Kim. "Selective inhibition of methanogens for the improvement of biohydrogen production in microbial electrolysis cells." *International Journal of Hydrogen Energy* 35, no. 24 (2010): 13379–13386.

Du, Zhuwei, Haoran Li, and Tingyue Gu. "A state of the art review on microbial fuel cells: a promising technology for wastewater treatment and bioenergy." *Biotechnology Advances* 25, no. 5 (2007): 464–482.

Fan, Yanzhen, Hongqiang Hu, and Hong Liu. "Enhanced Coulombic efficiency and power density of air-cathode microbial fuel cells with an improved cell configuration." *Journal of Power Sources* 171, no. 2 (2007): 348–354.

Franks, Ashley E., Kelly P. Nevin, Hongfei Jia, Mounir Izallalen, Trevor L. Woodard, and Derek R. Lovley. "Novel strategy for three-dimensional real-time imaging of microbial fuel cell communities: monitoring the inhibitory effects of proton accumulation within the anode biofilm." *Energy & Environmental Science* 2, no. 1 (2009): 113–119.

Freguia, Stefano, Korneel Rabaey, Zhiguo Yuan, and Jürg Keller. "Sequential anode–cathode configuration improves cathodic oxygen reduction and effluent quality of microbial fuel cells." *Water Research* 42, no. 6–7 (2008): 1387–1396.

Gorby, Yuri A., Svetlana Yanina, Jeffrey S. McLean, Kevin M. Rosso, Dianne Moyles, Alice Dohnalkova, Terry J. Beveridge et al. "Electrically conductive bacterial nanowires produced by Shewanella oneidensis strain MR-1 and other microorganisms." *Proceedings of the National Academy of Sciences* 103, no. 30 (2006): 11358–11363.

Harnisch, Falk, and Stefano Freguia. "A basic tutorial on cyclic voltammetry for the investigation of electroactive microbial biofilms." *Chemistry–An Asian Journal* 7, no. 3 (2012): 466–475.

Ieropoulos, Ioannis A., John Greenman, Chris Melhuish, and John Hart. "Comparative study of three types of microbial fuel cell." *Enzyme and Microbial Technology* 37, no. 2 (2005): 238–245.

Ki, D., J. Park, J. Lee, and K. Yoo. "Microbial diversity and population dynamics of activated sludge microbial communities participating in electricity generation in microbial fuel cells." *Water Science and Technology* 58, no. 11 (2008): 2195–2201.

Kim, Byung Hong, In Seop Chang, and Geoffrey M. Gadd. "Challenges in microbial fuel cell development and operation." *Applied Microbiology and Biotechnology* 76, no. 3 (2007): 485–494.

Kim, Hyung Joo, Hyung Soo Park, Moon Sik Hyun, In Seop Chang, Mia Kim, and Byung Hong Kim. "A mediator-less microbial fuel cell using a metal reducing bacterium, *Shewanella putrefaciens*." *Enzyme and Microbial Technology* 30, no. 2 (2002): 145–152.

Liu, Gang, Yu Tao, Ya Zhang, Maarten Lut, Willem-Jan Knibbe, Paul van der Wielen, Wentso Liu, Gertjan Medema, and Walter van der Meer. "Hotspots for selected metal elements and microbes accumulation and the corresponding water quality deterioration potential in an unchlorinated drinking water distribution system." *Water Research* 124 (2017): 435–445.

Liu, Hong, and Bruce E. Logan. "Electricity generation using an air-cathode single chamber microbial fuel cell in the presence and absence of a proton exchange membrane." *Environmental Science & Technology* 38, no. 14 (2004): 4040–4046.

Logan, Bruce E. *Microbial Fuel Cells*. John Wiley & Sons (2008).

Lovley, Derek R., and Kelly P. Nevin. "Electricity production with electricigens." In *Bioenergy*, pp. 293–306. Washington, DC, USA: ASM Press (2008).

Min, Booki, and Bruce E. Logan. "Continuous electricity generation from domestic wastewater and organic substrates in a flat plate microbial fuel cell." *Environmental Science & Technology* 38, no. 21 (2004): 5809–5814.

Mohan, Y., S. Manoj Muthu Kumar, and D. Das. "Electricity generation using microbial fuel cells." *International Journal of Hydrogen Energy* 33, no. 1 (2008): 423–426.

Niessen, Juliane, Falk Harnisch, Miriam Rosenbaum, Uwe Schröder, and Fritz Scholz. "Heat treated soil as convenient and versatile source of bacterial communities for microbial electricity generation." *Electrochemistry Communications* 8, no. 5 (2006): 869–873.

Park, Doo Hyun, and J. Gregory Zeikus. "Improved fuel cell and electrode designs for producing electricity from microbial degradation." *Biotechnology and Bioengineering* 81, no. 3 (2003): 348–355.

Park, Hyung Soo, Byung Hong Kim, Hyo Suk Kim, Hyung Joo Kim, Gwang Tae Kim, Mia Kim, In Seop Chang, Yong Keun Park, and Hyo Ihl Chang. "A novel electrochemically active and Fe (III)-reducing bacterium phylogenetically related to *Clostridium butyricum* isolated from a microbial fuel cell." *Anaerobe* 7, no. 6 (2001): 297–306.

Pham, Cuong Anh, Sung Je Jung, Nguyet Thu Phung, Jiyoung Lee, In Seop Chang, Byung Hong Kim, Hana Yi, and Jongsik Chun. "A novel electrochemically active and Fe (III)-reducing bacterium phylogenetically related to *Aeromonas hydrophila*, isolated from a microbial fuel cell." *FEMS Microbiology Letters* 223, no. 1 (2003): 129–134.

Ping, Qingyun, Barak Cohen, Carlos Dosoretz, and Zhen He. "Long-term investigation of fouling of cation and anion exchange membranes in microbial desalination cells." *Desalination* 325 (2013): 48–55.

Prasad, D., S. Arun, M. Murugesan, S. Padmanaban, R. S. Satyanarayanan, Sheela Berchmans, and V. Yegnaraman. "Direct electron transfer with yeast cells and construction of a mediatorless microbial fuel cell." *Biosensors and Bioelectronics* 22, no. 11 (2007): 2604–2610.

Rabaey, Korneel, and René A. Rozendal. "Microbial electrosynthesis—revisiting the electrical route for microbial production." *Nature Reviews Microbiology* 8, no. 10 (2010): 706–716.

Rabaey, Korneel, Jorge Rodríguez, Linda L. Blackall, Jurg Keller, Pamela Gross, Damien Batstone, Willy Verstraete, and Kenneth H. Nealson. "Microbial ecology meets electrochemistry: electricity-driven and driving communities." *The ISME Journal* 1, no. 1 (2007): 9–18.

Rabaey, Korneel, Nico Boon, Monica Höfte, and Willy Verstraete. "Microbial phenazine production enhances electron transfer in biofuel cells." *Environmental Science & Technology* 39, no. 9 (2005): 3401–3408.

Rabaey, Korneel, Nico Boon, Vincent Denef, Marc Verhaege, Monica Höfte, and Willy Verstraete. "Bacteria produce and use redox mediators for electron transfer in microbial fuel cells." In *228th National Meeting of the American-Chemical-Society*. 2004.

Rinaldi, Antonio, Barbara Mecheri, Virgilio Garavaglia, Silvia Licoccia, Paolo Di Nardo, and Enrico Traversa. "Engineering materials and biology to boost performance of microbial fuel cells: a critical review." *Energy & Environmental Science* 1, no. 4 (2008): 417–429.

Rinaldi, Antonio, Barbara Mecheri, Virgilio Garavaglia, Silvia Licoccia, Paolo Di Nardo, and Enrico Traversa. "Engineering materials and biology to boost performance of microbial fuel cells: a critical review." *Energy & Environmental Science* 1, no. 4 (2008): 417–429.

Ringeisen, Bradley R., C. M. Othon, Xingjia Wu, D. B. Krizman, M. M. Darfler, J. J. Anders, and P. K. Wu. "Biological laser printing (BioLP) for high resolution cell deposition." In *Cell and Organ Printing*, pp. 81–93. Springer, Dordrecht (2010).

Ringeisen, Bradley R., Ricky Ray, and Brenda Little. "A miniature microbial fuel cell operating with an aerobic anode chamber." *Journal of Power Sources* 165, no. 2 (2007): 591–597.

Rismani-Yazdi, Hamid, Ann D. Christy, Burk A. Dehority, Mark Morrison, Zhongtang Yu, and Olli H. Tuovinen. "Electricity generation from cellulose by rumen microorganisms in microbial fuel cells." *Biotechnology and Bioengineering* 97, no. 6 (2007): 1398–1407.

Rismani-Yazdi, Hamid, Sarah M. Carver, Ann D. Christy, and Olli H. Tuovinen. "Cathodic limitations in microbial fuel cells: an overview." *Journal of Power Sources* 180, no. 2 (2008): 683–694.

Sharma, Vinay, and P. P. Kundu. "Biocatalysts in microbial fuel cells." *Enzyme and Microbial Technology* 47, no. 5 (2010): 179–188.

Taskan, Ergin, Bestamin Ozkaya, and Halil Hasar. "Combination of a novel electrode material and artificial mediators to enhance power generation in an MFC." *Water Science and Technology* 71, no. 3 (2015): 320–328.

Wang, Aijie, Lihong Liu, Dan Sun, Nanqi Ren, and Duu-Jong Lee. "Isolation of Fe (III)-reducing fermentative bacterium *Bacteroides* sp. W7 in the anode suspension of a microbial electrolysis cell (MEC)." *International Journal of Hydrogen Energy* 35, no. 7 (2010): 3178–3182.

Wang, Junfeng, Xinshan Song, Yuhui Wang, Junhong Bai, Manjie Li, Guoqiang Dong, Fanda Lin, Yanfeng Lv, and Denghua Yan. "Bioenergy generation and rhizodegradation as affected by microbial community distribution in a coupled constructed wetland-microbial fuel cell system associated with three macrophytes." *Science of the Total Environment* 607 (2017): 53–62.

Wei, Jincheng, Peng Liang, and Xia Huang. "Recent progress in electrodes for microbial fuel cells." *Bioresource Technology* 102, no. 20 (2011): 9335–9344.

You, Shijie, Qingliang Zhao, Jinna Zhang, Junqiu Jiang, and Shiqi Zhao. "A microbial fuel cell using permanganate as the cathodic electron acceptor." *Journal of Power Sources* 162, no. 2 (2006): 1409–1415.

Zhang, Yifeng, Booki Min, Liping Huang, and Irini Angelidaki. "Generation of electricity and analysis of microbial communities in wheat straw biomass-powered microbial fuel cells." *Applied and Environmental Microbiology* 75, no. 11 (2009): 3389–3395.

Zhao, Feng, Nelli Rahunen, John R. Varcoe, Alexander J. Roberts, Claudio Avignone-Rossa, Alfred E. Thumser, and Robert CT Slade. "Factors affecting the performance of microbial fuel cells for sulfur pollutants removal." *Biosensors and Bioelectronics* 24, no. 7 (2009): 1931–1936.

Zuo, Yi, Defeng Xing, John M. Regan, and Bruce E. Logan. "Isolation of the exoelectrogenic bacterium Ochrobactrum anthropi YZ-1 by using a U-tube microbial fuel cell." *Applied and Environmental Microbiology* 74, no. 10 (2008): 3130–3137.

Index

H

I

L

M

N

O

P

Q

R

S

T

U

X

Y

Z

Taylor & Francis eBooks

www.taylorfrancis.com

A single destination for eBooks from Taylor & Francis with increased functionality and an improved user experience to meet the needs of our customers.

90,000+ eBooks of award-winning academic content in Humanities, Social Science, Science, Technology, Engineering, and Medical written by a global network of editors and authors.

TAYLOR & FRANCIS EBOOKS OFFERS:

A streamlined experience for our library customers

A single point of discovery for all of our eBook content

Improved search and discovery of content at both book and chapter level

REQUEST A FREE TRIAL

support@taylorfrancis.com

Printed in the United States
by Baker & Taylor Publisher Services